Disclaimer

The publisher of this book is by no way associated with the National Institute of Standards and Technology (NIST). The NIST did not publish this book. It was published by 50 page publications under the public domain license.

50 Page Publications.

Book Title: Analysis of Pipeline Steel Corrosion Data

Book Author: Richard E. Ricker

Book Abstract: Currently, the U.S. has over 3.7 million kilometers (2.3 million miles) of pipelines crossing the country transporting natural gas and hazardous liquids from sources such as wells, refineries, and ports to customers. It is estimated that almost 2/3 of the energy consumed in the U.S. passes through a pipeline at some point between its origin and the point of consumption and pipelines account for about 20% of the total mass-distance that oil and natural gas are transported [1,2]. Clearly, the maintenance of an uninterrupted energy supply to the public requires the operation of these pipelines in such a manner that corrosion does not result in an unscheduled interruption to the flow of these energetic materials to the nation, as occurred recently in Alaska [3]. This task is accomplished by pipeline operating companies, who follow standards, codes, and practices set out by variety of regulatory agencies, industrial consortia, and standards developing organizations. the Pipeline Standards developing Organizations coordination Council (PSDOCC) coordinates the activities of these groups and the Department of Transportations Office of Pipeline Safety (OPS) is the main regulatory agency with final responsibility over this system of codes and practices [2].

Citation: NIST Interagency/Internal Report (NISTIR) - NISTIR 7415

Keyword: corrosion;pipline;pitting;statistics;steel

NISTIR 7415

Analysis of Pipeline Steel Corrosion Data From NBS (NIST) Studies Conducted Between 1922-1940 and Relevance to Pipeline Management

Richard E. Ricker
Materials Performance Group
Metallurgy Division
Materials Science and Engineering Laboratory
National Institute of Standards and Technology
Gaithersburg, MD 20899

National Institute of Standards and Technology
Technology Administration, U.S. Department of Commerce

NISTIR 7415

Analysis of Pipeline Steel Corrosion Data From NBS (NIST) Studies Conducted Between 1922-1940 and Relevance to Pipeline Management

Richard E. Ricker
Materials Performance Group
Metallurgy Division
Materials Science and Engineering Laboratory
National Institute of Standards and Technology
Gaithersburg, MD 20899

May 2, 2007

U.S. Department of Commerce
Carlos M. Gutierrez, Secretary

Technology Administration
Robert Cresanti, Under Secretary of Commerce for Technology

National Institute of Standards and Technology

Analysis of Pipeline Steel Corrosion Data From NBS (NIST) Studies Conducted Between 1922-1940 and Relevance to Pipeline Management

Executive Summary

Between 1911 and 1984, the National Bureau of Standards (NBS) conducted a large number of corrosion studies that included the measurement of corrosion damage to samples exposed to real-world environments. One of these studies was an investigation conducted between 1922 and 1940 into the corrosion of bare steel and wrought iron pipes buried underground at 47 different sites representing different soil types across the Unites States. At the start of this study, very little was known about the corrosion of ferrous alloys underground. The objectives of this study were to determine (i) if coatings would be required to prevent corrosion, and (ii) if soil properties could be used to predict corrosion and determine when coatings would be required. While this study determined very quickly that coatings would be required for some soils, it found that the results were so divergent that even generalities based on this data must be drawn with care. The investigators concluded that so many diverse factors influence corrosion rates underground that planning of proper tests and interpretation of the results were matters of considerable difficulty and that quantitative interpretations or extrapolations could be done "only in approximate fashion" and attempted only in the "restricted area" of the tests until more complete information is available.

Following the passage of the Pipeline Safety Improvement Act in 2002 and at the urging of the pipeline industry, the Office of Pipeline Safety of the U.S. Department of Transportation approached the National Institute of Standards and Technology (NBS became NIST in 1988) and requested that the data from this study be reexamined to determine if the information handling and analysis capabilities of modern computers and software could enable the extraction of more meaningful information from these data. This report is a summary of the resulting investigations.

The data from the original NBS studies were analyzed using a variety of commercially available software packages for statistical analysis. The emphasis was on identifying trends in the data that could be later exploited in the development of an empirical model for predicting the range of expected corrosion behavior for any given set of soil chemistry and conditions. A large number of issues were identified with this corrosion dataset, but given the limited knowledge of corrosion and statistical analysis at the time the study was conducted, these shortcomings are not surprising and many of these were recognized by the investigators before the study was concluded. However, it is important to keep in mind that complete soil data is provided for less than half of the sites in this study. In agreement with the initial study, it was concluded that any differences in the corrosion behavior of the alloys could not be resolved due to the scatter in the results from the environmental factors and no significant difference could be determined between alloys. Linear regression and curve fitting of the corrosion damage measurements against the measured soil composition and properties found some weak trends. These trends improved with multiple regression, and empirical equations representing the performance of the samples in the tests were developed with uncertainty estimates. The uncertainties in these empirical models for the corrosion data were large, and extrapolation beyond the parameter space or exposure times of these experiments will create additional uncertainties.

It is concluded that equations for the estimation of corrosion damage distributions and rates can be developed from these data, but these models will always have relatively large uncertainties that will limit their utility. These uncertainties result from the scatter in the measurements due to annual, seasonal, and sample position dependent variations at the burial sites. The data indicate that more complete datasets with soil property measurements reflecting the properties of the soil and ground water directly in contact with the sample from statistically designed experiments would greatly reduce this scatter and enable more representative predictions.

Analysis of Pipeline Steel Corrosion Data From NBS (NIST) Studies Conducted Between 1922-1940 and Relevance to Pipeline Management

Richard E. Ricker

Materials Performance Group
Metallurgy Division
Materials Science and Engineering Laboratory
National Institute of Standards and Technology
Technology Administration
U.S. Department of Commerce
Gaithersburg, MD 20899

I. Introduction

Currently, the U.S. has over 3.7 million kilometers (2.3 million miles) of pipelines crossing the country transporting natural gas and hazardous liquids from sources such as wells, refineries, and ports to customers. It is estimated that almost 2/3 of the energy consumed in the U.S. passes through a pipeline at some point between its origin and the point of consumption and that pipelines account for about 20 % of the total mass-distance that oil and natural gas are transported [1, 2]. Clearly, the maintenance of an uninterrupted energy supply to the public requires the operation of these pipelines in such a manner that corrosion does not result in an unscheduled interruption to the flow of these energetic materials to the nation, as occurred recently in Alaska [3]. This task is accomplished by pipeline operating companies, who follow standards, codes, and practices set out by a variety of regulatory agencies, industrial consortia, and standards developing organizations. The Pipeline Standards Developing Organizations Coordination Council (PSDOCC) coordinates the activities of these groups, and the Department of Transportation's Office of Pipeline Safety (OPS) is the main regulatory agency with final responsibility over this system of codes and practices [2].

Following pipeline accidents in Carlsbad, NM [4] and Bellingham, WA [5], the U.S. Congress passed the Pipeline Safety Improvement Act of 2002 (PSIA). The objective of this act was to improve public safety by stimulating improvements in pipeline technologies, regulations, and standards. This act resulted in the formation of the PSIA Coordination Council, which communicates and coordinates pipeline relevant research in four government agencies: The Department of Energy, The Department of Transportation, The Department of Interior, and The Department of Commerce. This project is a result of this collaboration. The objectives of this project were to (1) reexamine the original NBS underground bare pipe corrosion studies to determine if the results from this study could be used to develop better empirical models for prediction of bare pipe corrosion rates and (2) to seek new in-sights that could lead to the development of pipeline external corrosion prediction models, or soil corrosivity indexes, that could be used in the future for computer-aided pipeline management.

Since the mileage of existing pipelines greatly exceeds that of new construction, the average age of the U.S. pipeline infrastructure is increasing steadily [6]. Penetration of the pipeline wall

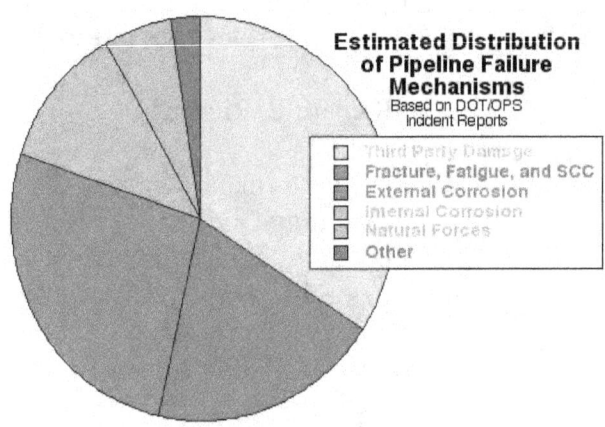

Figure 1 - Attributed failure mechanisms for reported pipeline failures.

as a result of corrosion of the external surface is responsible for a significant portion of pipeline failures as shown in Figure 1 [6]. Assuming that corrosion rates are greater than zero, this means that the threat of corrosion-induced failures is actually increasing steadily each year. Considering this, it is surprising that this industry has been able to actually reduce or hold failure rates constant over recent years as shown in Figure 2 [6]. This feat has been accomplished through the accumulation of pipeline operation experience, and improvements in technologies including inspection, repair, coating, and information technologies. This industry openly shares their experience through a number of different consortia and standards developing organizations. As a result, the practices, codes, and standards developed reflect this experience and evolve as pipelines age and new technologies are developed. This has enabled this industry to make improvements and repairs before failures actually occur. This industry can be expected to make further improvements as better inspection, repair, and information technologies are developed that allow this industry to monitor, measure, and track changes in their pipelines to even greater resolutions and detail.

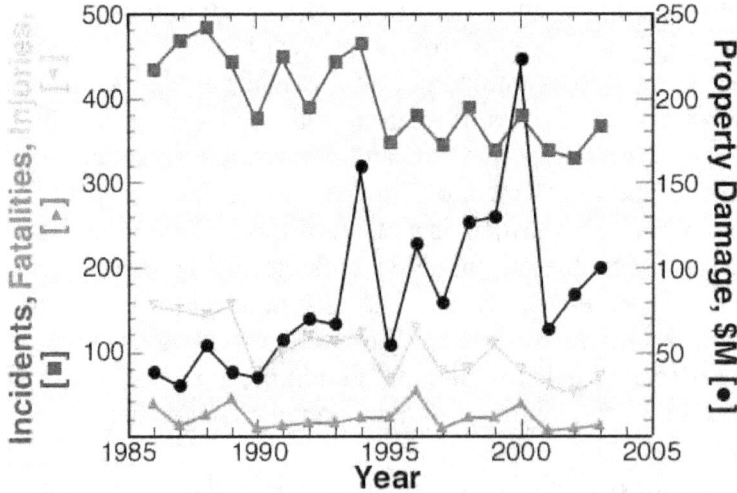

Figure 2 - Statistics from the Office of Pipeline Safety on pipeline accidents

It is conceivable that in the future pipeline operators will have computer systems that provide data on every meter of the pipelines in their system at their fingertips. Ideally, one of the parameters for each increment of the pipe will be an indicator of the corrosivity of the local environment to the steel of the pipeline wall. This measure may be based on sensor readings or estimated from measures of soil properties and chemistry. This parameter will give the operator information on how long the steel pipe should be able to contain its contents without catastrophic failure should the coating and/or cathodic protection systems fail. Ideally, this parameter will help the operator schedule inspections and plan shutdowns for repairs so that the energy supply is uninterrupted. The quality of the decisions that operators will make based on this parameter will depend on the accuracy or the uncertainty in the estimate of this parameter. Currently, operators are required to use the same "corrosion allowance" for all corrosion rate based decision-making unless they can prove that a lower corrosion rate can be expected.

The value used for the current corrosion allowance was determined by analysis of underground corrosion measurements taken by the National Bureau of Standards (NBS) during a study conducted between 1920 and 1947 [7, 8]. The same value is to be used for all soils and underground pipeline environments without regard for the specific soil chemistry of each site and local conditions. There is a provision for exceptions when an operator can demonstrate that lower rates can be expected for a particular section. This is a conservative approach at present that grows more conservative as information and other technologies improve. Advances in computers, sensors, chemical property measurements, and computer modeling of chemical reactions and transport can be expected to make this an overly conservative approach in the near future. These emerging technologies will make the acquisition and manipulation of increasingly detailed information on increasingly smaller increments of a pipeline possible. The first step toward accomplishing this next level of corrosion allowance determination should be the establishment of a link between some measurable property of the pipeline soil environment and the resulting corrosion rate. There are essentially three different approaches that can be taken to establish this link: (1) empirical correlations to actual measurements of corrosion damage in steel pipes exposed to representative soils, (2) development of laboratory measurements and models for estimation, and (3) detailed computer models with valid assumptions for rate determining processes. Each of these different approaches has advantages and drawbacks, but all three will require verification with actual data from exposure tests on samples in representative soil environments. Therefore, the first of these is the logical starting point especially since it will help one identify the critical issues for the other two.

Information and data on the corrosion behavior of steels in underground environment is rather limited, and many studies into underground corrosion rely on the data from the studies conducted by NBS between 1920 and 1947 [7, 8]. The data from these studies have been used for underground corrosion decision making over a wide range of fields from underground utilities to nuclear waste disposal. These studies actually began in 1911 when Congress asked NBS to conduct studies into electrolysis failures caused by the operation of electric streetcars. In conducting this study it was noted that very little was known about how steels and other metals should corrode underground in the absence of induced electric currents induced from the operation of streetcars. As this study was nearing completion, it was noted that the emerging pipeline industry had a critical need for this type of information. As a result, a workshop was held at NBS with participants from industry and an underground corrosion research program was proposed. The Department of Agriculture was asked to identify locations with representative soils and to participate in the characterization of the soils at the sites. Industry was asked to provide samples and to participate in sample burials, removals, and inspections. Workshops were convened at regular intervals to keep everyone updated and were attended by corrosion experts from all over the world. This study lead to a large number of similar studies of corrosion in real world conditions and eventually into the development of related laboratory research programs in corrosion measurement methods at NBS that evolved over time into the present programs in the Metallurgy Division of NIST.

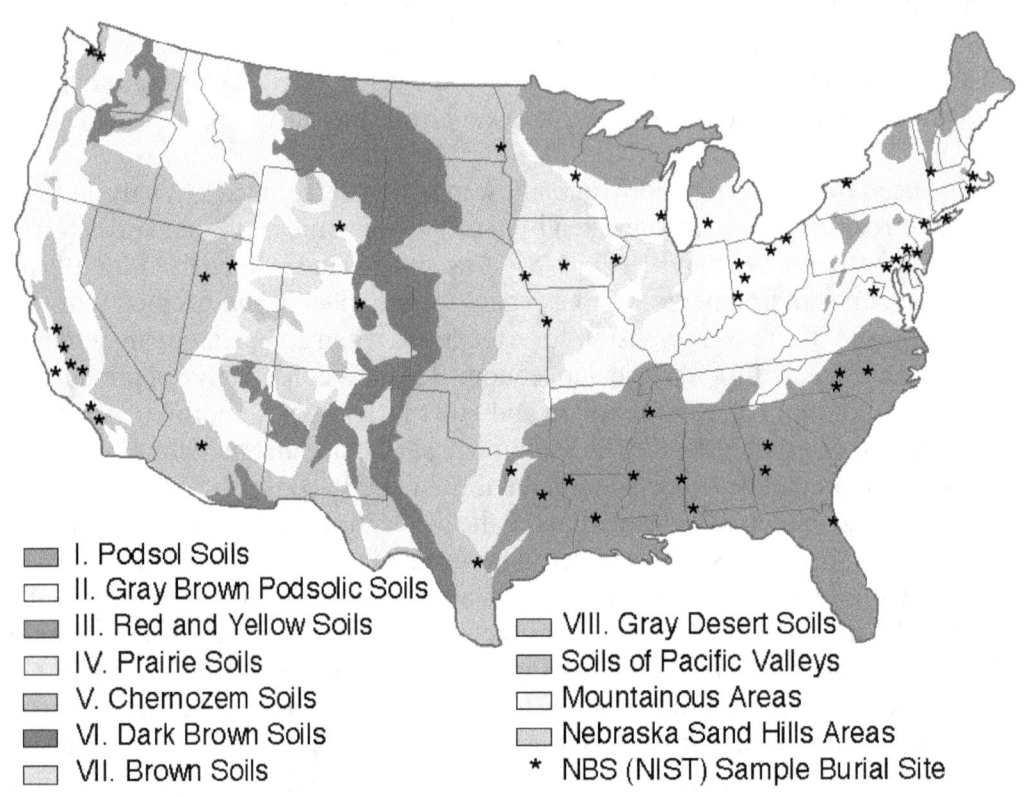

Figure 3 – Map of the US showing the locations of the burial sites and the 8 major soil groups identified in the study.

The original NBS bare pipe underground corrosion studies incorporated 47 sites across the United States as shown in Figure 3. In this figure, the 8 basic soil types are identified as they were in the 1957 summary report. Since the 1957 report was prepared, the Department of Agriculture has subdivided soil groups and currently lists 13 major soil groups in the United States [9]. Detailed soil maps with these updated classifications can be obtained from the Department of Agriculture [9]. Samples were retrieved from sites at periodic intervals, with the last samples removed between 12 and 17 years after burial depending on the site. Figure 4 is a photograph of the samples from this study laid out in the NBS laboratory for examination. The bare pipe corrosion study was the first of a long series of studies of corrosion in real world situations conducted by NBS [7, 8].

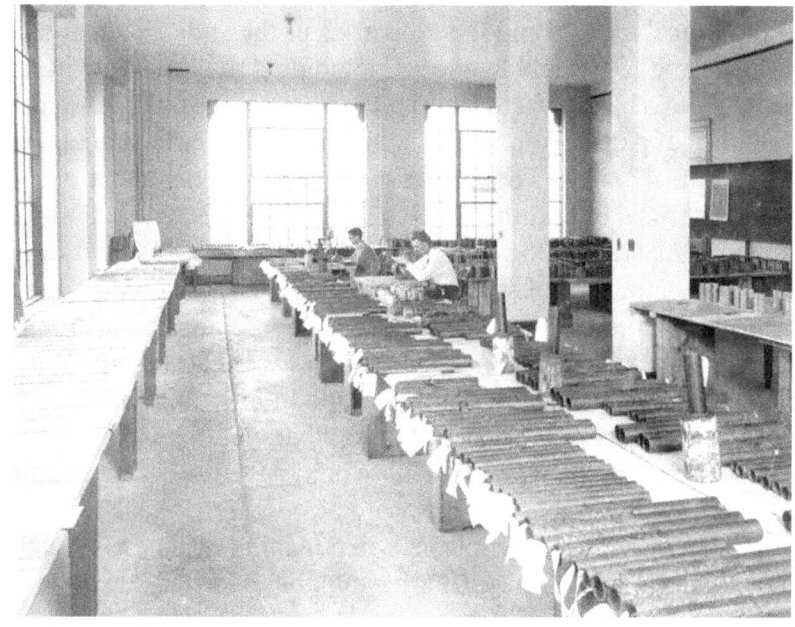

Figure 4 – Samples being examined in the laboratories at NBS.

When NBS began the underground corrosion studies of bare steel pipes (1922-24), the Department of Agriculture identified sites for the placement of coupons and conducted soil surveys to characterize the soils. Soil samples were then analyzed by NBS to determine the composition and properties of the soils at the sites. Soil surveys and taxonomy were new concepts just being developed in the 1920s [9]. The soil property measurements and chemical analyses were also state-of-the-art for the time the study was conducted. Statistical analysis was not a well developed and appreciated part of metrology when these studies were designed. The NBS underground corrosion studies have been criticized for the poor statistical design of the experiments including neglecting the distribution of the samples at the sites [10]. That is, soil horizons, while mostly parallel to the surface, frequently vary in depth even over the short distance of a burial trench. As a result, samples from opposite ends of the same trench could be exposed to different conditions. A well designed experiment for statistical analysis would have the samples distributed in the trenches in a manner that avoids this spatial bias. In addition, seasonal and annual bias can result from variations in starting dates, exposure times (fractional years), and the use of average annual data for conditions rather than measurements.

The original NBS underground pipeline corrosion study appears to have attempted to mimic the pipeline burial conditions and practices of the day for each location (e.g. burial depth varies with location). In this manner, the results would represent the uncertainties inherent in these practices and conditions rather than just fundamental information on the influence of soil chemistry and properties on corrosion rates. For example, the annual rainfall given for each site is actually the average annual rainfall for the location closest to the burial site with rainfall data and not an average for the actual site or the years of burial. At the time of the study, this was the only kind of information that would be available to a pipeline operator and remote sensing, recording, or monitoring was not to be a consideration for decades. In addition, many soil properties were measured in the laboratory rather than in the field. Removing soils from the ground will alter the activity of important species such as water, carbon dioxide and oxygen and alter the activity of biological species and the properties of the soils. The impact of these factors on pH was recognized by the 1950s and Romanoff attempted to correct the pH values [8]. In addition, cost appears to have been a factor during the studies limiting site selection, sample layout, and examination. Of particular concern is the fact that chemical analyses were conducted on soils from only 26 of the 47 sites. Today, statistical analysis considerations would dictate that all sites should be characterized or that the 26 sites should be selected at random. However, the sites with soil chemistry data are the 26 with the lowest measured electrical resistivity; and therefore, the highest concentrations of soluble salts. These issues should not be considered the fault of the original investigators because their importance in obtaining data for statistical analysis was not fully appreciated at the time these studies were designed. It appears that the original study decided to emulate the buried pipeline conditions, soil characterization data, and the associated uncertainties inherent in the information that would be available to pipeline operators. This would mean that the resulting data would have greater scatter than might result for more controlled conditions, but this scatter would represent the "real-world" uncertainties that pipeline regulators and operators would confront when making decisions. Including this scatter in the data insured the data would be representative and that decisions made would be conservative for the prevailing conditions of the day. Decades later, this seems to be an overly conservative approach that inhibits statistical analysis, interpretation, and the development of performance prediction models.

In the near future, information technologies, sensors, and global information systems (GIS) will make it possible to characterize or even monitor environmental chemistries and soil properties at closely spaced intervals along a pipeline. Not only will pipeline operators have larger quantities of better soil characterization measurements, they will have better tools for manipulating and interpreting this information. Computer aided monitoring, data manipulation, and operation decision-making is becoming standard practice. All of these possibilities were not even a consideration in the 1920s when the original NBS study was initiated. In fact, when the study initiated it was not even clear that coatings would be required to protect pipes from corrosion. Today, cathodic protection and coatings are used extensively. One of the most significant impacts of the original study may have been to determine that coating would be required and to stimulate coatings research and development. Coatings for pipeline protection and pipeline coating technologies are still a major area of research and development today [11].

For the analyses of this study, it will be assumed that the soils removed from the trench were used for backfill and that the chemical and physical properties given in the summary reports accurately represent the soil in physical contact with the samples at each site [7, 8]. Since there is no information on uncertainty or variability in the reports [7, 8], thorough homogenization to these precise values must be assumed. Similarly, without seasonal data, it must be assumed that there are no seasonal variations in conditions and corrosion rates that would allow for the exact dates of placement and retrieval and fractional years of exposure to have an influence on the results. Also, it must be assumed that the years of burial were typical years so that the annual rainfall, temperature, and other soil characteristics are properly represented by the data. The additional uncertainty imposed by these assumptions should be kept in mind along with the conclusion in the 1957 summary report that the statistical variability in the data was too great to make reliable predictions possible [8].

II. Burial Site Characterization

The term soil is usually used to describe any of the naturally occurring loose collections of solid particles found on the surface of the earth that support the growth of plants [9]. This includes the inorganic minerals, organic species, liquids, and gasses found in these aggregates. According to a strict interpretation of this definition, a soil only extends as deep as the roots of the plants or other organic species that grow in the soil. Today it is understood that soils are alive with organic species of all types and sizes [12]. Pipelines are typically buried below the levels where these organic species are plentiful and special backfill free of organics may be used rather than the dirt and soils removed from the trench. This does not mean that microorganisms will not influence the corrosion rate of a pipeline. Sulfate-reducing bacteria have long been known to stimulate corrosion of steel pipelines in anaerobic environments with sufficient nutrient content [13, 14]. In addition, biological activity both in the soils above the pipeline as well as those above the surface of the soil may have a dramatic influence on the water, oxygen, and carbon dioxide content at the burial depth of the pipe. However, these factors were not quantified for examination in the original NBS study and cannot be considered here. In addition, it will be assumed that the backfill was the soil removed from the trench and that the chemical analysis of the soil at each site provided in the summary reports covering these studies presents a reasonable estimate of the chemical environment the samples experienced during these exposures [7, 8].

Soils are composed of essentially four features (1) mineral particulates, (2) organic matter from surface and subsurface plant and animal life, (3) groundwater containing soluble salts, and (4) gases. The particulate matter found in soils is usually small particles of the minerals found in the nearby rock formations that were produced from these formations over millions of years of weathering and the decomposition products produced when these minerals react with air and water. In either case, most of the particles making up a soil are insoluble minerals, as most soluble species have been removed over the millions of years of weathering. The solubility of these minerals may vary with pH, and if they do so, they will tend to buffer the pH of the groundwater, but assuming no significant changes in pH with time, we can assume for a first order approximation that these minerals behave as inert solids. Soils are placed into categories as sands, clays, silts, or loams based on the size distribution of these particles as shown in Figure 5. The potential influence of organic matter either living and excreting potentially corrosive compounds or decaying and producing potentially corrosive conditions locally should not be ignored in a thorough life prediction scheme, but information on these conditions were not collected with the data of these studies. It is also important to realize that the properties assigned to a site may change over time due to human, animal, or plant activity, but there is also no information on these types of changes occurring at the burial sites.

It should be kept in mind that the objectives of the original NBS study were (1) to determine if bare pipe could be used in some or all soils and (2) to determine if measures of soil characteristics could be used to predict the corrosivity of soils and enable better pipeline decision-making and management. To accomplish the second objective, one might want to include all of the natural range of variability that could be expected for normal pipeline burial practices of the day, since it is the extreme rates that will produce failures. At the time these studies were conducted, data were manipulated and analyzed manually and a single soil sample

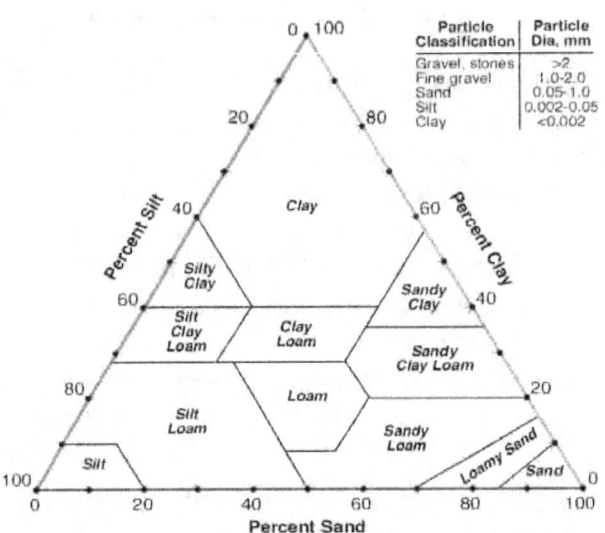

Figure 5 - Ternary diagram describing soil types by characteristic particle sizes.

might be used to represent the soils and exposure conditions for a considerable length of pipeline. Since at that time statistical tools for addressing these issues were very limited, it is logical that one would want to include the complete natural range of actual conditions that a single set of soil property data might be used to represent. Any attempt to control or limit this natural variability might be viewed as producing data less representative of "real-world" industrial practice since it would not include the entire range of conditions and rates expected for a soil with the properties indicated by the soil sample.

The sites for burial of the samples were identified by the Department of Agriculture and they were selected to represent the different types of soils and conditions that could be found in the U.S. The sites were identified by number, location, and soil type as shown in Table A1 (Appendix A contains the tables of site descriptions and measured characteristics). Table A2 lists the 26 different parameters used to identify or represent the properties of the soils found at the burial sites along with the units used in the original reports and the current SI equivalent units with the conversion factors used for this study. Some measures were arbitrary ratings such as fair, good, and poor for site internal drainage, some were measures of soil properties, and some were taken from locally available data such as average annual rainfall and ground temperature. Most of the soil properties were measured for all 47 sites (Tables A3 and A5), but the chemistry of water extract was determined for only the 26 sites with the lowest resistivity measurements (Tables A4 and A6). Table A7 contains the complete descriptions of the soil horizons and depths for all of the sites used for this study.

The relationships between electrical conductivity of the medium or electrolyte and corrosion behavior have been the subject of much debate and some research [15-18]. The conductivity of an electrolyte is the product of the concentration of charge carrying species and their mobility. The ionic bonding of the insoluble mineral particles will prevent conduction through these particles. Therefore, electrical conduction will be restricted to the solutions in the pore spaces around the particles and the conductivity of the water-saturated soil is a measure of the soluble salts present in the soil to form ions in the water, the volume fraction of pore space, and the mobility of the charge carrying ions. Fortunately, most ions other than the hydrogen and hydroxide ion have similar mobilities in aqueous solutions. So, conductivity is simply an estimate of the total ion content of the solution surrounding the particles of soil. In general, corrosion rates are observed to increase with the conductivity of a soil. Increasing the conductivity of the electrolyte enables greater separation of the cathodic and anodic half-cell reactions. It also reduces the range of potential differences that are possible between different sites on the surface of the sample. Escalante et al. [15-18] examined the effects of conductivity, temperature, and mass transport in soils and found that while lower conductivities tended to result in lower average corrosion (mass loss) rates, they also resulted in a greater range of variations in corrosion rates across the surface. This would result in increased pitting ratios that could result in wall penetration rates equivalent to those of more corrosive environments with lower pitting ratios. That is, while lower conductivities (higher resistivities) tend to result in lower overall corrosion rates, it also makes it easier for corrosion to localize to a small spot or region of the surface and form pits. This was observed and reported in the original NBS studies and identified in the summary reports as a major factor contributing to the scatter in the data that made reliable corrosion predictions difficult [7, 8].

The first step in analyzing the data from the NBS bare pipe underground corrosion study was to plot cumulative distribution functions (CDF) for the measurements characterizing the properties of the soils at the sites (Table A3) as shown in Figure 6. In these figures, the x-axis is the measured property and the y-axis is the fraction or percentage of sites having this value for the property or less. In this manner the CDF goes from 0 to 1 or 0 % to 100 % over the measured range for the variable. The slope of the CDF is the more familiar probability density function (PDF) or the fraction of sites within some bandwidth of the value given on the x-axis (i.e., density). A log-normal distribution was used for all measures of chemical concentrations or measures that would relate to chemical reactivity, as chemical reaction kinetics typically vary with the log of the activity or concentration of the reaction species (Table A4 and Figure 7)[19]. The exception to this is pH, as it is a log scale.

A standard score (Z) was calculated for each characteristic (i) at each site (j) according to the relationship

$$Z_{ij} = \frac{x_{ij} - \mu_x}{\sigma_x} \qquad (1)$$

where x_{ij} is the measured value of the characteristic for the normal distributions and the logarithm of the measurement for the log-normal distributions and μ_x and σ_x are the corresponding mean and standard deviation values of this property. Converting the measurements to a standardized variable (Tables A5 and A6) produces scores for analysis that are without units and can be compared on the same graph with the same scale without bias. Conversion from the Z-score back to original units is simple matter of applying the mean and standard deviation given in Tables A5 and A6 through Equation (1) above.

In addition to the measured physical and chemical properties of the sites, the depth and nature of the soil horizons (horizontal layers or strata) were qualitatively characterized, and these are included along with the depth that the samples were buried in Table A7. This table is included to illustrate the complex nature of the soils at the sites and to demonstrate how the behavior of samples from one part of a site to another could vary if depth of the horizons varied.

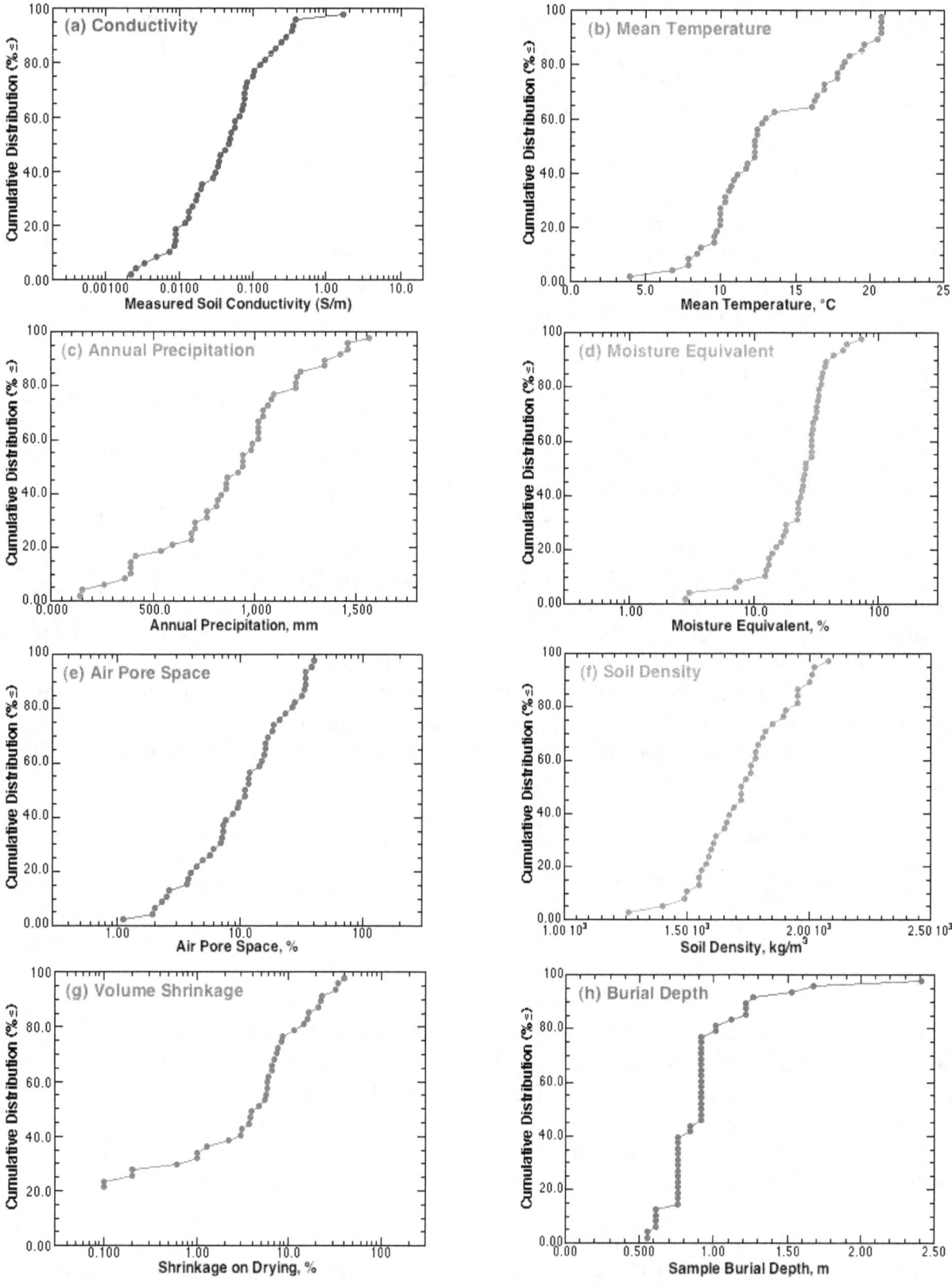

Figure 6 – Cumulative distributions functions for the measured properties of the soils.

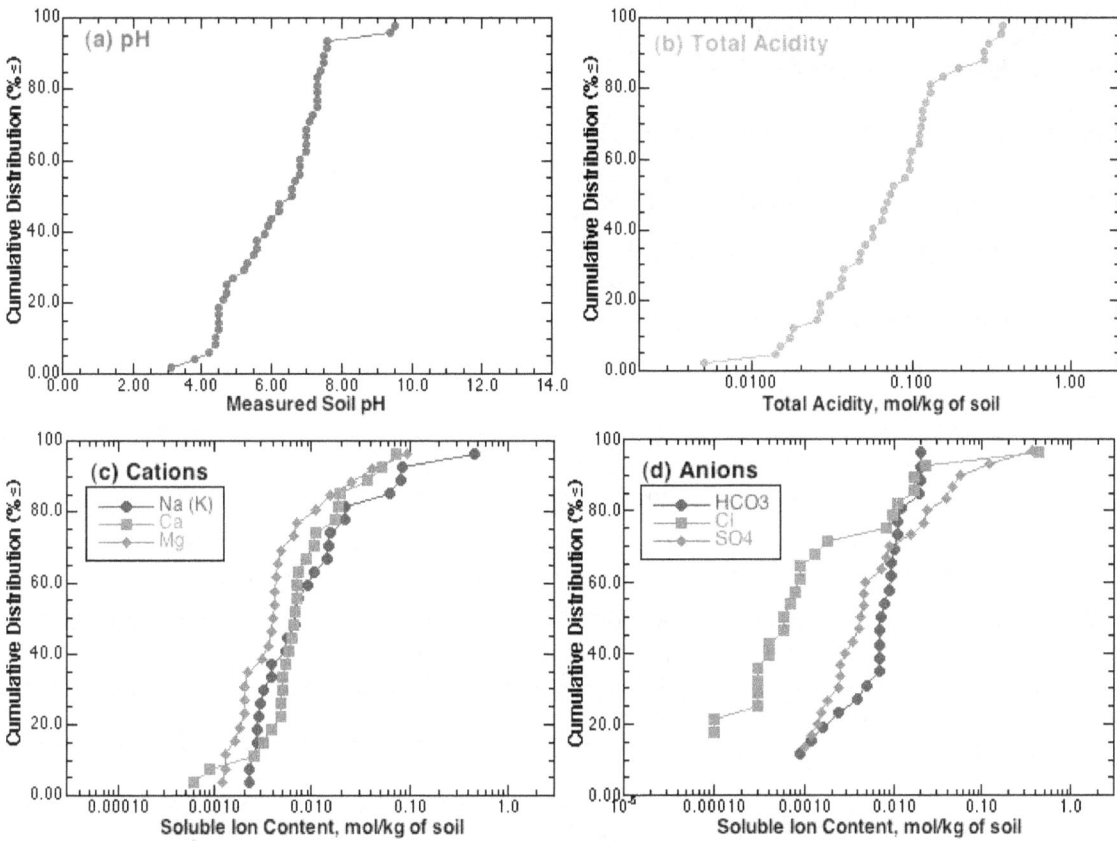

Figure 7 – Cumulative distribution functions for the concentrations of soluble chemical species in soils and total acidity: (a) pH, (b) total acidity, (c) cations, and (d) anions.

III. Corrosion Damage Characterization

Appendix B contains the tables of corrosion damage measurements. In addition, Table A2 includes measured units and conversions used for the measures of corrosion damage along with those for the site characterization parameters. Corrosion damage was characterized by measuring two factors: (1) mass change and (2) pipe wall thinning. The mass change was measured after the corrosion products were removed in a manner such that the underlying metal would remain intact. The descaling procedures used in the studies are described in the 1945 report [7]. In addition to the average mass change for two samples, the average of the deepest penetration into the wall of two pipes was reported. This results in two measures of damage for the exposure: (1) mass loss and (2) corrosion penetration. These two can be converted to rates by dividing by the exposure time. This calculation assumes that the corrosion rates are effectively constant over the exposure time. Of course, mass loss can be converted to an average penetration rate using the density of the metal as shown in Table A2. The ratio of the maximum measured penetration to the average penetration calculated from mass loss is the pitting ratio, which is a measure of the propensity of the exposure environment to cause local variations in the corrosion rate over the surface of a sample (a pit being a high corrosion rate at a small spot). In this study, essentially three corrosion response variables were studied: (1) the mass loss rate

(MLR), (2) the corrosion penetration rate (CPR), and (3) the pitting ratio. As with the environmental variables, the units and conversion for these measures are given in Table A2.

Eight different types of samples were buried at each sight with 6 sets of duplicates for periodic retrieval. The samples were provided as nominal 1.5 inch and 3 inch pipe (38.1 mm and 114.3 mm). Table B1 identifies sample size and alloy composition by the single letter used to identify each sample type: "a," "b," "e," "y," "B," "K," "M," and "Y". The alloys and microstructures of these samples almost certainly deviate significantly from those available today primarily due to the dramatic improvements in processing that has reduced slag inclusions and mill scale. Apparently, in an effort to accurately represent the conditions of actual buried pipeline, no special effort was put into cleaning sample surface and removing mill scale beyond that required to remove oils and allow sealing of the ends with caps. Mill scale and inclusions are typically noble with respect to the Fe of the metal and the presence of these phases on the surfaces will stimulate cathodic activity enabling higher corrosion rates that might be localized to the region around these phases depending on the nature (conductivity) of the surrounding soil. In addition, the graphite phases in the microstructure will also tend to act as sites for cathodic (reduction) reactions, and the finer more controlled microstructures available today will reduce the tendency of these features to localize corrosion. However, without hard data on these differences and their impact, it must be assumed that these alloys represent the range of alloys used in pipelines past and present.

The logistics and cost of maintaining exact year exposure increments in these studies outweighed the desire for data from identical exposures. If there is a seasonal variation in the corrosion rate, it will contribute to the unquantifiable scatter in these measurements that cannot be explained since there are insufficient data on the dates of burial and retrieval. This is significant because it is possible that a site may have rainy and dry seasons such that almost all of the "annual corrosion damage" occurs in one season. In this case, an exposure of 1.25 y could have twice the damage of an exposure of 1.0 y. Seasonal variations are frequently observed in real world exposure tests, but these cannot be addressed with the current dataset.

As pointed out in the section above on site characterization, the corrosion damage to the samples was quantified by measuring the change in the mass of the samples over the burial period and by measuring the maximum depth of wall penetration in the samples. The mass loss was measured after removal of corrosion products and the descaling techniques are described in the 1945 summary report. The mass loss was reported as the average mass loss per unit area (oz/ft^2) for two samples of each of 8 different types of ferrous pipeline alloys. These measurements converted to the current SI units for mass loss (g/m^2) are presented in Table B2. The exposure times were given with the mass loss data and converting these to mass loss rates in grams per meter squared per day ($g/m^2/d$) results in the data presented in Table B3. Similarly, the maximum corrosion penetration measurements were presented as the average maximum depth of penetration (mils) for two samples. These measurements converted to SI units (mm) are given in Table B4 and after conversion to penetration rates (mm/y) in Table B5.

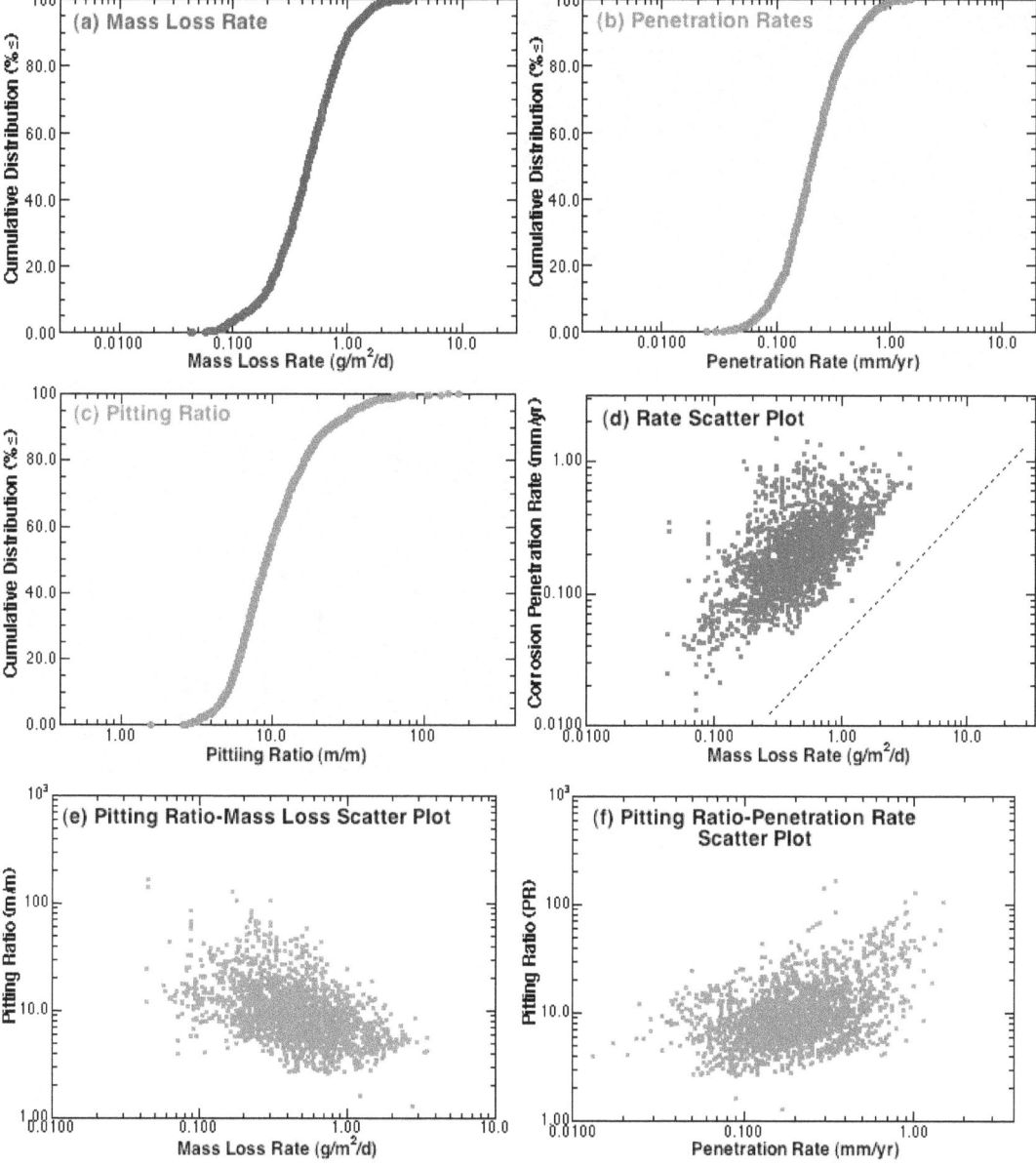

Figure 8 – Cumulative distribution functions for the corrosion mass loss (a) and penetration rates (b) and the normalized ratio of these rates or pitting ratio (c). Scatter plots examining the relationships between these measures of corrosion damage rates: (d) penetration v. mass loss, (e) pitting ratio v. mass loss, and (f) pitting ratio v. penetration.

The cumulative distribution functions for the corrosion mass loss rates, penetration rates, and pitting ratios are shown in Figures 8(a) through 8(c). These figures show that combining the measurements from all of the sites results in a smooth and symmetric sigmoidal curve when plotted as the log of the rate or ratio indicating that log-normal distributions can be used to represent these data. Plotting these measures against each other with log scales as in Figures 8(d) through 8(f), shows that there is no clear trend relating these measures and that the data form ellipsoidal scatter plots. Segregating the data into subgroups based on alloy type results in the cumulative distribution functions shown in Figure 9(a) through 9(c). By examining these

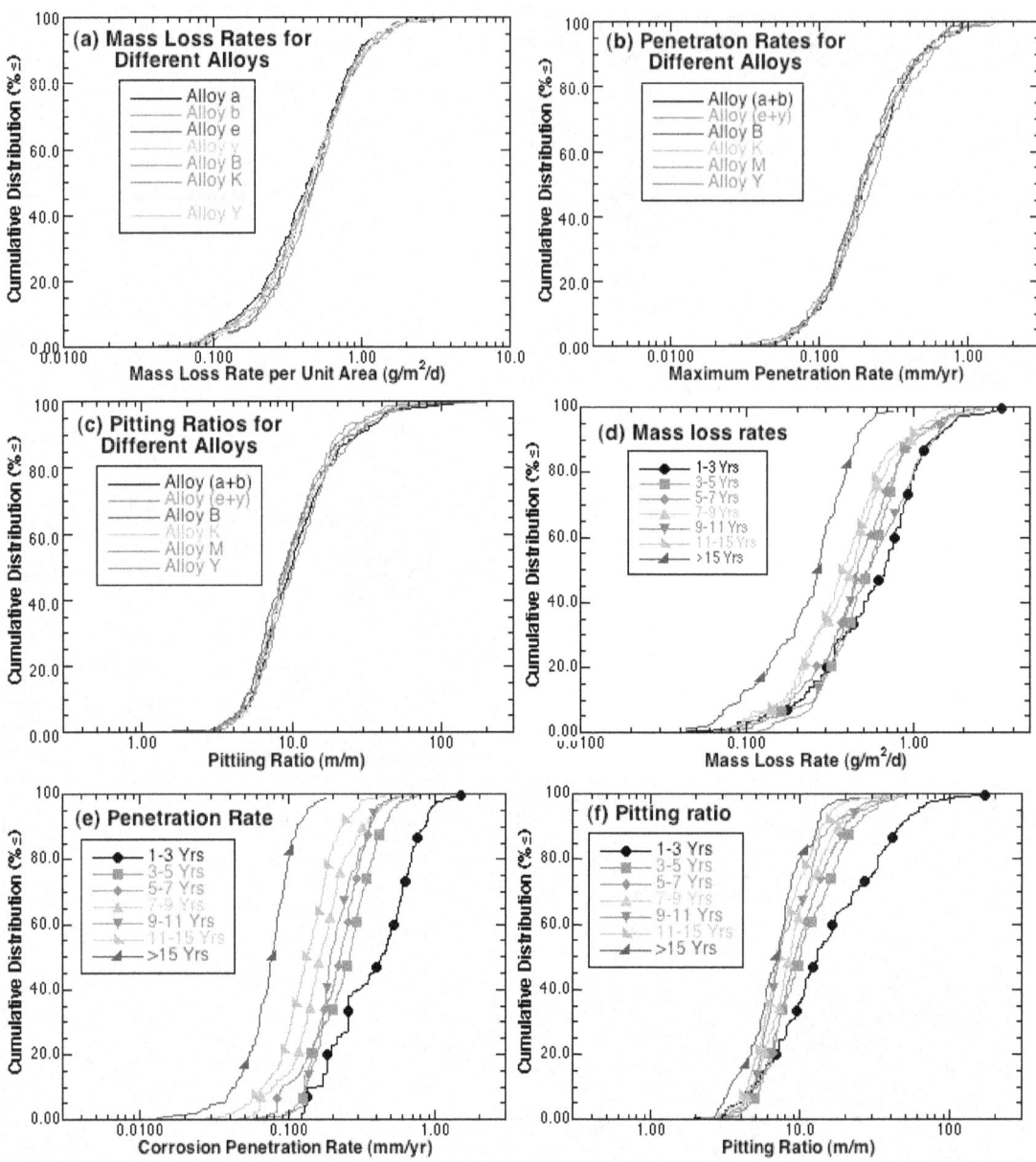

Figure 9 – Cumulative distribution functions examining the effects of alloy composition and exposure time on the measurement: (a) Mass loss rates for different alloys, (b) Corrosion penetration rates for different alloys, (c) Pitting ratios for different alloys, (d) Mass loss rates for different retrieval periods, (e) Corrosion penetration rates for different retrieval periods, and (f) Pitting ratios for different retrieval periods.

figures, it is clear that scatter in the measurements resulting from the exposure variables and the natural stochastic nature of underground corrosion overwhelms any differences due to alloy type for this range of alloy compositions. This is not uncommon for steels [20, 21]. Therefore, subsequent analyses will assume that measurements from these alloys can be considered to be from the same alloy and analyzed as such to add numbers and statistical weight to the trends. On the other hand, breaking the measurements into subsets according to the length of time that the samples were underground indicates that both mass loss rates and corrosion penetration rates decreased with the time that the samples were buried in the ground. In addition, the pitting ratios also tended to decrease with exposure time. This is an important observation that will be discussed later in this report.

Since the exposed surface area of the samples could influence the observed maxima in penetration depth and sample types "a," "b," "e," and "y" had almost exactly half the exposed area of sample types "B," "K," "M," and "Y," subsequent analyses of the corrosion penetration rates were done by taking the maximum reported for sample types "a" or "b" as a single measurement and similarly the maximum for sample types "e" and "y" as a single measurement. As a result, there are 6 measurements of maximum corrosion penetration rate per site and retrieval while there are 8 measurement of mass loss rate.

To briefly illustrate the range of variations among the different exposure sites for these measures of corrosion damage, sites representing the extreme maximum and minimum for the mean and range of these 3 corrosion measures are plotted in Figure 10. Figure 10(a) illustrates the motivation for this and similar studies of corrosion damage rates. This figure shows that the maximum mass loss rate observed on any sample at Site 6 is at least an order of magnitude lower than lowest rate observed for any sample at Site 23. Clearly, these corrosion rates depend strongly on the characteristics of these sites and identifying the characteristics that can be used to reliable identify which range of behavior a pipeline will exhibit will enable better management decision making. However, Figure 10(b) illustrates one of the main problems for accomplishing this objective. This figure shows the maximum penetrations observed for the samples at the same two sites shown in Figure 10(a). The corrosion penetration rate distributions for these two sites overlap. Since these two sites represent the extremes in the mean log penetration rate, all of the other sites fall between these two and also overlap. This clearly illustrate the trend for the sites with lower mass loss rates to have a greater range of corrosion rates over the surface area of the samples resulting in more localized high rates or pitting. The sites representing those with the greatest and smallest range in the three measures of corrosion damage are shown in Figures 10(d) through 10(f). Again, the mass loss rate measurements indicate over an order of magnitude difference with the variation being in proportion with the differences in the means creating two nearly parallel lines on the log scale. On the other hand, the corrosion penetration rates and pitting ratios shown in Figure 10(d) and 10(e) do not form smooth continuous curves, but show irregular "jumps" in the curves indicating that samples above and below these "discontinuities" experience different conditions or that stochastic variations in processes resulted in the nucleation or creation of highly corrosive conditions. The discontinuity, rather than a gradual slope change, suggests that there is a threshold or nucleation event that separates the behavior of the pit from that of the remainder of the surface.

IV. Environment-Corrosion Rate Relationships

The relationship between the three measures of corrosion damage and the quantitative variables describing the properties and chemistry of the soils at the sites were explored by plotting the standard score for the variables at the site against the corrosion damage measure and performing linear regression on the measurements using commercially available curve fitting software. Some of the better results from this regression process are illustrated in Figure 11. By

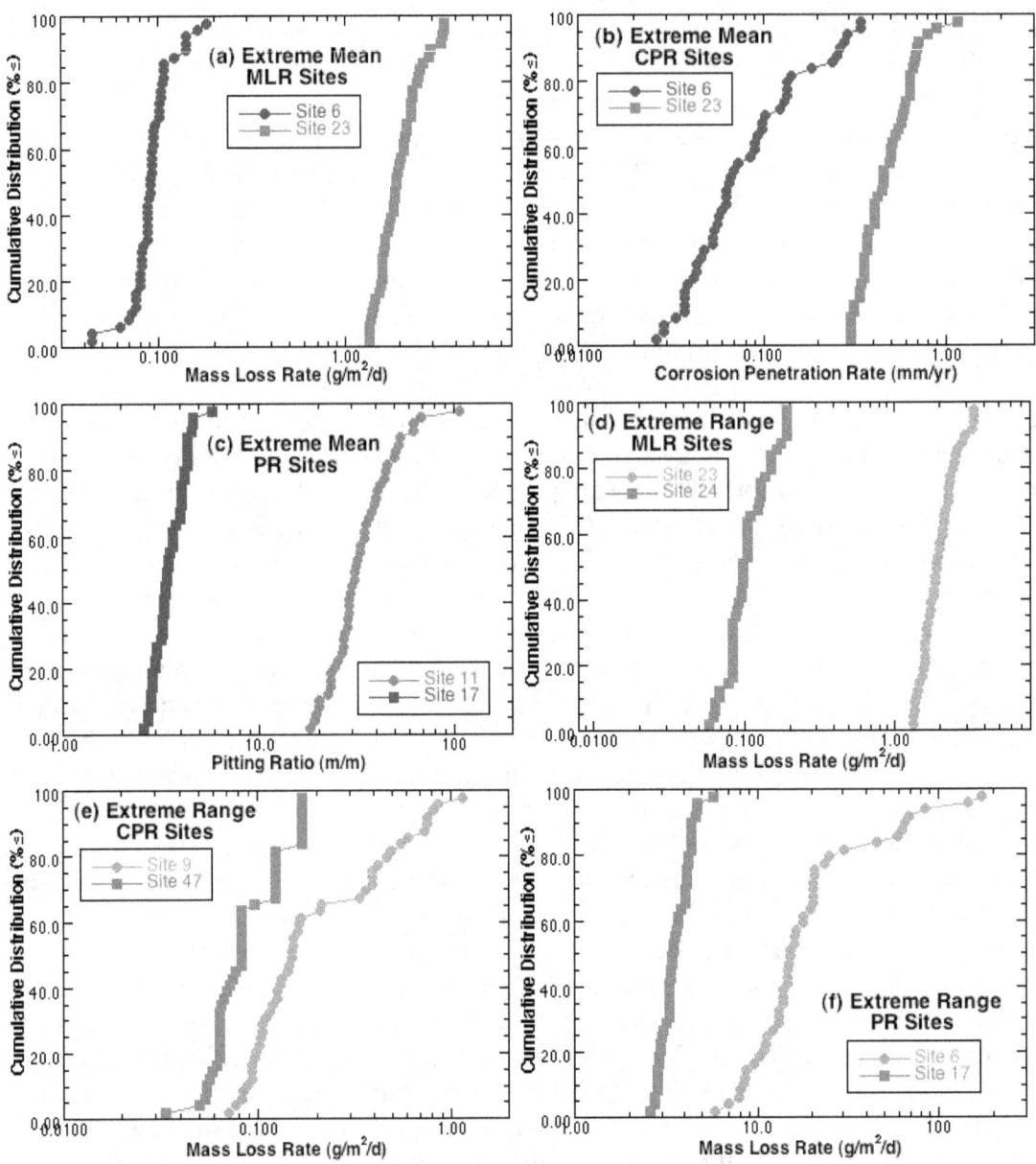

Figure 10 – Cumulative distribution functions for sites illustrating the range of behavior observed. Sites exhibiting minimum and maximum (a) mean mass loss rate, (b) mean corrosion penetration rate, and (c) mean pitting ratio and sites with the minimum and maximum range for (d) mass loss rates, (e) corrosion penetration rates, and (f) pitting ratios.

examining this figure, it can be seen that none of the variables exhibited well-defined trends with any of the corrosion measures. The correlation coefficient for a curve fit is the ratio of the unexplained variation to the explained variation; and therefore, is 0 when there is no indicated relationship between the parameters and has a magnitude of 1 when the curve fit can describe the exact location of every point. The correlation coefficients for the fit of these site characterization variables to the corrosion damage measures are given in Table C1. This process was repeated taking all of the samples for all of the sites as individual measurement points and the correlation

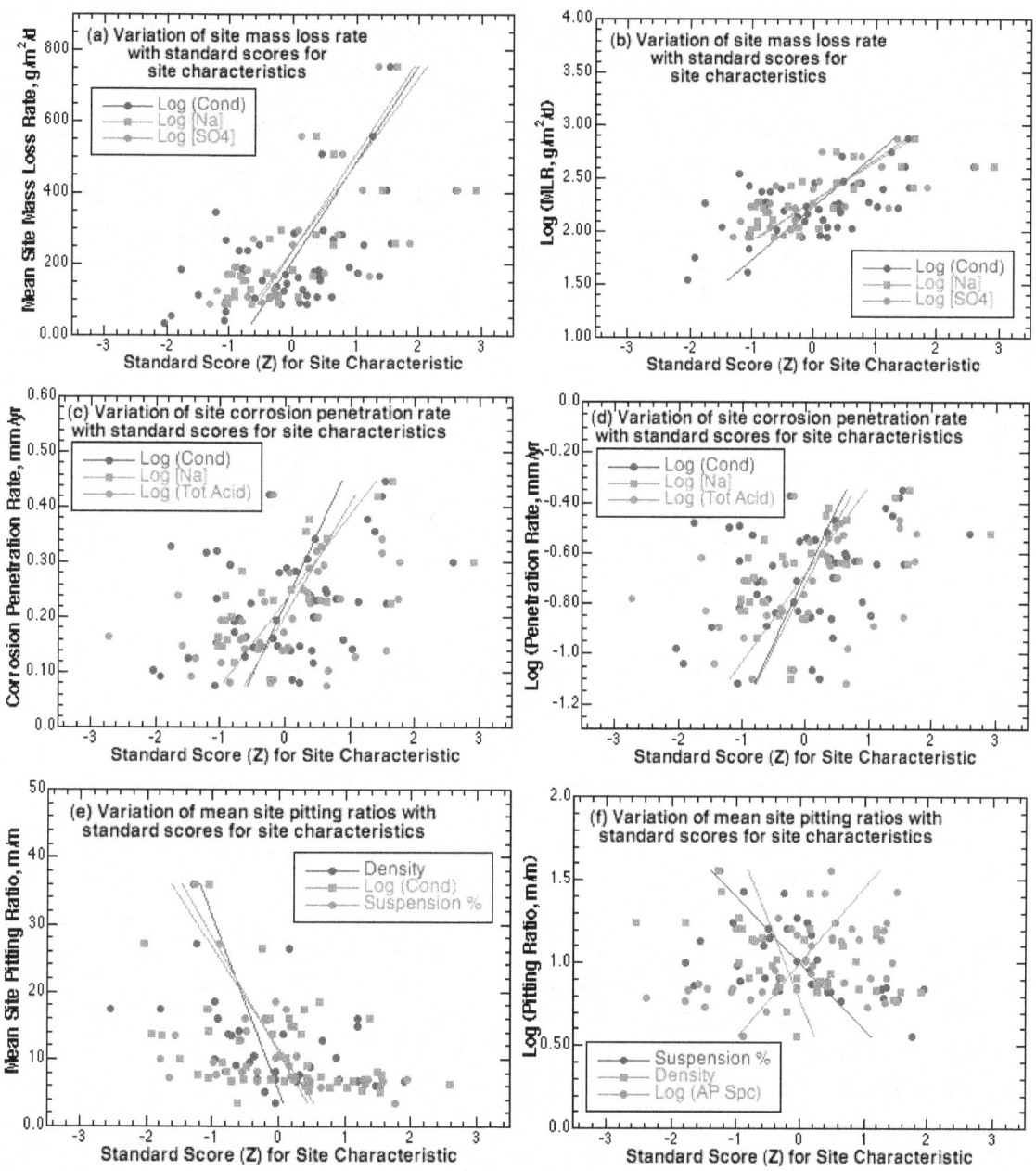

Figure 11 – Linear regression results for fitting (a) site mean mass loss rates (MLR), (b) Log (MLR), (c) site mean corrosion penetration rates (CPR), (d) Log (CPR), (e) site mean pitting ratios (PR) and (f) log (PR) for selected site characterization parameters (see Tables C1 and C2).

coefficient for these fits are also shown in Table C1. By examining this table it can be seen that the best fit was found for the mass loss rate (log) as one might expect after examination of Figures 8-10. The site characterization variables with the highest correlation coefficient were the concentrations of the ions Na^{-1} and SO_4^{-2}. The fits to the corrosion penetration rate were slightly lower with the same site characterizations variable resulting in the highest correlation coefficients. However, the highest correlation coefficient observed for any of the single variable regression fits was 0.714 that is not a particularly good fit as shown by the Na^{-1} line of fit in Figure 11(a).

After examining linear regression fits, multiple regression analyses were performed on the site averages for the corrosion damage measures. Given the wide range of possible combinations and the number of variables, experimenting with scientifically logical and derivative fits proved to be a very time consuming and, given the poor quality of most fits, disappointing process. However, a scheme was developed and followed for the evolution of a fit. This scheme results in a completely empirical fit in that there is essentially no scientific consideration given to the selection of the variable used in the derivation of the fit other than it was selected for measurement in the original study. Basically, each corrosion damage measure was fit against each of the site characterization variables taking one at a time (single linear regression). Then the variable that produced the best fit was used in 2-term regression model using all of the remaining site characterization variables taking one at a time. This process was repeated for 3, 4, and 5 term multiple regression models. At each step, the site characterization variables yielding the second and third best correlation coefficient were examined in place of the best fit to insure that the best fitting variable was selected.

The exception to this process was soil conductivity. Since the conductivity of the soil is a measure of the total ion content of the soils, it is a measure of the combined concentrations of all soluble ions. Since ion chemistry was measured for only 26 of the 47 sites, using any of the ions concentrations in the fit significantly reduces the number of points being fit. This was considered undesirable particularly for the early terms in the process. In addition, when Na^{-1} fit well with a measure of corrosion, the anions Cl^{-1} and SO_4^{-2} also showed higher correlations making it unclear which was the more important. Using conductivity, at last in the first term, allows for representation of ion concentrations without forcing an empirical selection of an ion that may not be important in determining the rate as much as simply varying with the ions that do matter. It should be kept in mind that these are totally empirical fits and may not even indicate the important variable as many of these site characterization variables are interrelated, and this empirical variable selection process may result in the selection of a variable that varies with an important property, but was measured or represents the causative property better for the sites than the measures used to quantify that property.

The result of this term-by-term multiple regression fitting process are given in Tables C3 and C4 for fitting the site average mass loss rate and corrosion penetration rates respectively. The predictive capability of these models is illustrated graphically in Figure 12. The correlation coefficient for multiple regression fits for prediction of the site average mass loss rate was 0.942 and 0.956 for 5 and 6 terms, respectively. Similarly, the fits for the corrosion penetration rate yielded correlation coefficients of 0.860 and 0.891 for 5 and 6 terms. These empirical fits allow for estimation of the mean, average, or expected value for a site given the properties used in the calculation. The scatter in the fits allows for the estimation of the scatter that should be observed at a site characterized by the variables. That is, the uncertainty in the curve fit is illustrated graphically in Figure 12 by the dashed lines for the confidence interval.

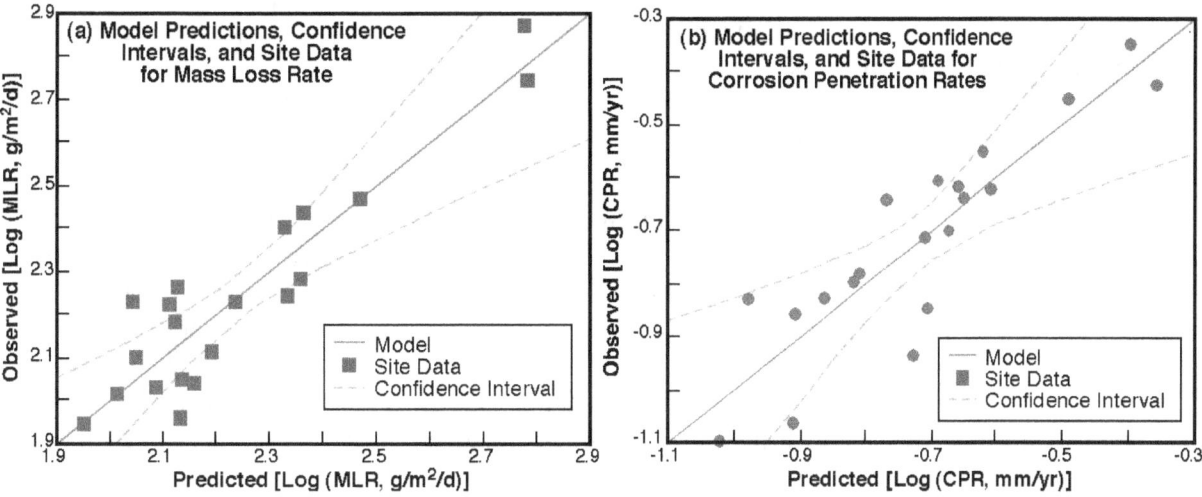

Figure 12 – Multiple regression modeling results for mass loss rates (a) and corrosion penetration rates (b) as a function of site characteristics.

V. Variation of Rate with Exposure Time

As shown in Figure 9, the mass loss rate and corrosion penetration rates tended to decrease with exposure time. A decrease in the corrosion rate with time is not unexpected as there are a number of different kinetic rate models that would predict such a trend [19-21]. First, if corrosion products build up on the surface and this inhibits the transport of reactants to or from the surface, the corrosion rate will decrease as this layer grows thicker if mass transport through this layer is rate limiting. Similarly, if there is a cathodic reactant that is being consumed by corrosion and it is being depleted from the surrounding environment, a slow decline in the corrosion rate with time is to be expected. In either case, the measured (average) corrosion damage rate will decrease with increasing exposure time. The behavior of the measured corrosion damage as a function of time will indicate the mechanism responsible for the declining rate. Fitting corrosion damage to a power law equation of the form

$$y = at^n \quad (2)$$

where y is the measure of corrosion damage and t is the exposure time and the constants a and n are determined by the fitting process [7, 8, 20, 21]. In the case where corrosion damage is constant with respect to time, the exponent, n, will be one and in the case of a growing barrier film n will be 0.5 and other postulated rate limiting mechanisms may yield other values. An n-value greater than one would indicate that the corrosion rate increased with exposure time. While the nucleation of pitting after some incubation period longer than the first or second retrieval could result in n-values greater than 1, pit nucleation times are usually very much shorter than these exposure times and n-values between 0 and 1 are frequently observed for corrosion damage rates [21].

This time dependence was recognized in the original NBS studies and they examine this trend by linear regression of the equation

$$\log(y) = \log(a) + n\log(t) \quad (3)$$

with y equal to the two sample average maximum penetration for the exposure time (t). For this report, the measurements from each site for mass loss and corrosion penetration were fit to Equation (2) using commercial software for iterative non-linear curve fitting that uses a Levenberg-Marquardt algorithm for estimating successive iterations until the squares of the errors reach a minimum [22]. The use of a non-linear curve fitting routine allows for inclusion of the initial (zero exposure time, zero damage) data points in the curve fits that cannot be included in a linear regression of Equation (3). The results of these fits are shown in Figure 13 along with the results reported by in the NBS underground corrosion reports [7, 8] for linear regression per Equation (3). Figure 13(a) is a CDF for the fitting exponent (n) and Figure 13(b) is a PDF for this same parameter. By examining these figures, it can be seen that the corrosion penetration rate and the mass loss rate exhibit significantly different time dependences. That is, the corrosion penetration rate tends to slow to a much greater extent with exposure time than the mass loss rate. This indicates that the corrosion penetration rate is being limited by the mass transport of cathodic reactants or anodic products through the corrosion products building up at the pit while the rate limiting processes governing the mass loss rate and not facing the same restrictions. This also explains the trends shown in Figures 9(d)-9(f). Figures 13(c) and 13(d) show the correlation coefficients determined for the curve fits for the mass loss data and the corrosion penetrations respectively and these figures show that with the exception of two points, most of the correlation coefficients for mass loss were above 0.9 and above 0.8 for the penetration data. These figures also show that there is no clear trend in the correlation coefficients with the value determined for the power law exponent (n).

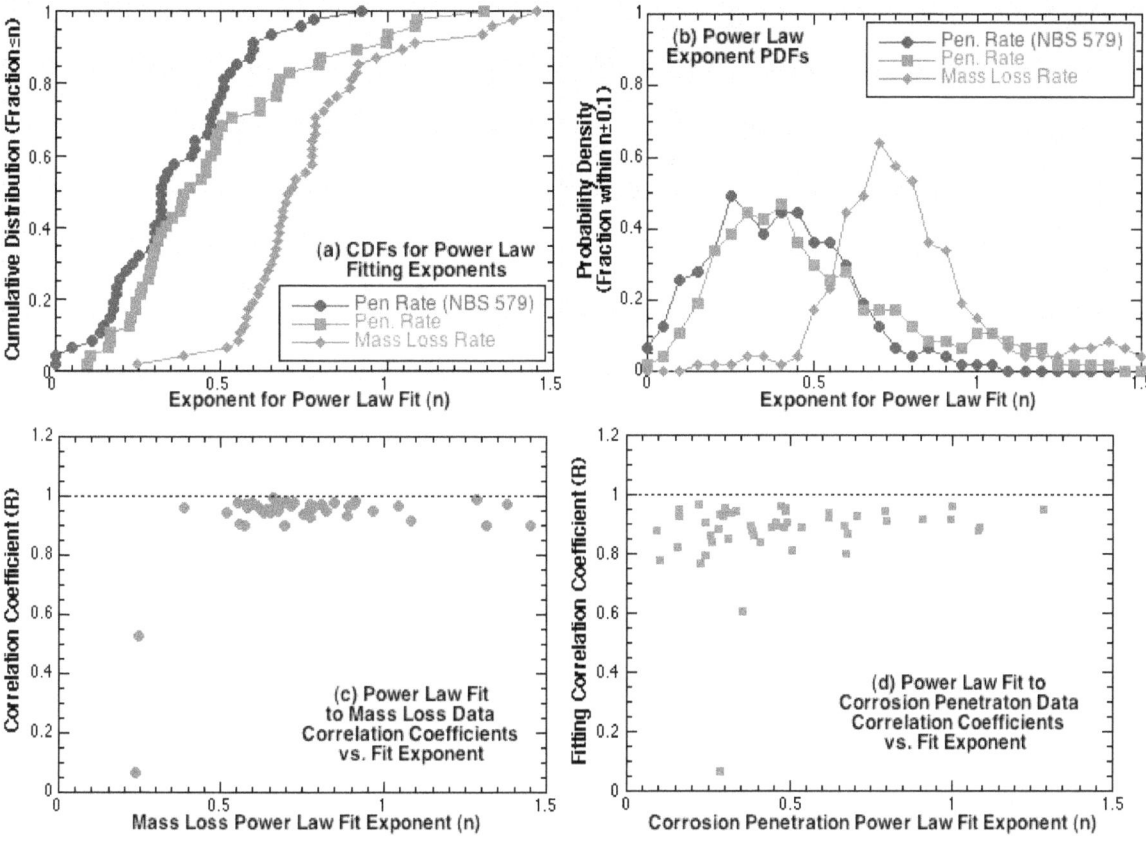

Figure 13 – Graphic presentation of the results of fitting mass loss and corrosion penetration measurements for each site to a power law equation: (a) cumulative distribution functions for exponents determine by fit, (b) probability distribution functions for exponents determined by fit, (c) variation of correlation coefficients for mass loss with exponent of fit, and (d) variation of the correlation coefficients for penetration with exponent of fit.

VI. Conclusions

After extensive examination and reexaminations of the data presented in the NBS studies of underground corrosion it is concluded that while equations for the estimation of corrosion damage distributions and rates can be developed from these data, that the scatter inherent in these models is considerable larger than it could be and that this will always limit the ability of predictions to be made from models based on this data. The scatter in these measurements is the result of the state-of-the-art at the time the study was conducted and the limitations of budget and size of the project. The data indicate that more complete datasets with soil property measurements reflecting the properties of the soil and ground water directly in contact with the samples including annual and seasonal variations and obtained with statistical analysis of the results considered during the design of the experimental program would greatly reduce this scatter and enable more representative predictions.

Acknowledgements

The author would like to express his gratitude to R. Smith and J. Merritt of the Office of Pipeline Safety for advice, help, and support. Also, the author would like to thank W. Leucke, D. Pitchure, and K. Synder for their review of this report and their numerous helpful comments and suggestions.

Disclaimer

While commercially available equipment and software were used for these studies, their use does not constitute an endorsement by the author or NIST nor should it be taken to imply that these are the best available for this purpose.

References

1. Dennis, S.M., *Improved Estimates of Ton-Miles.* Journal of Transportation and Statistics, 2005. 8(1): p. 23-44.
2. Mead, K.M., Actions Taken and Actions Needed to Improve Pipeline Safety, Office of Inspector General, U.S. Department of Transportation, CC-2004-055, Washington, DC, 2004.
3. Mufson, S., *Pipeline Closure Sends Oil Higher*, in *Washington Post*. 2006: Washington, DC.
4. NTSB, Natural Gas Pipeline Rupture and Fire Near Carlsbad, New Mexico August 19, 2000, Natonal Transportation Safety Board, Washington, DC, 2003.
5. NTSB, Pipeline Rupture and Subsequent Fire in Bellingham, Washington June 10, 1999, Natonal Transportation Safety Board, Washington, DC, 2002.
6. OPS, *Pipeline Statistics.* 2007, Office of Pipeline Safety, Pipeline and Hazardous Materials Administration, U.S. Department of Transportation.
7. Logan, K.H., Underground Corrosion, National Bureau of Standards (USA), Washington, DC, 1945.
8. Romanoff, M., Underground Corrosion, National Bureau of Standards (USA), NBS Circular 579, Gaithersburg, MD, 1957.
9. USDA, Soil Taxonomy A Basic System of Soil Classification for Making and Interpreting Soil Surveys, The Soil Survey Staff, Natural Resources Conservation Service, U.S. Dept. of Agriculture, Washington, DC, 1999.
10. Prasad, J.S. and T. Watmough, Evaluation of Literature on Buried Pipe Corrosion, IIT Research Institute, IITRI-B8093-3, Chicago, IL, 1967.
11. Ricker, R.E., Advanced Coatings R&D for Pipelines and Related Facilities, National Institute of Standards and Technology, NIST SP 1044, Gaithersburg, MD 20899, 2005.
12. Sudgen, A., R. Stone, and C. Ash, *Soils - The Final Frontier.* Science, 2004. 304(11 June): p. 1613.
13. Dexter, S.C., ed. *Biologically Induced Corrosion.* International Corrosion Conference Series, ed. S.C. Dexter. 1985, NACE International: Houston, TX.
14. von Wolozogen Kuhr, C.A.H. and L.S. Van der Vlugt, *Graphitization of Cast Iron as an Electro-biochemical Process in Anaerobic Soils.* Water, 1934. 18(16): p. 147-165.

15. Escalante, E., T. Oka, and U. Bertocci, *Effect of Oxygen Transport and Resistivity of the Environment on the Corrosion of Steel*, in *Mater. Res. Soc. Symp. Proc.* 1990. p. 303.
16. Escalante, E., T. Oka, and U. Bertocci, Effect of Oxygen Transport and Resistivity of the Environment on the Corrosion of Steel, National Instute of Standards and Technology (USA), Gaithersburg, MD, 1990.
17. Escalante, E., *Measuring the Corrosion of Metals in Soil*, in *Corrosion Testing and Evaluation: Silver Anniversary Volume*, R. Baboian and S.W. Dean, Editors. 1990, ASTM: Phil., PA. p. 112-124.
18. Escalante, E., *The Effect of Soil Resistivity and Soil Temperature on the Corrosion of Galvanically Coupled Metals in Soil*, in *Galvanic Corrosion*, H.P. Hack, Editor. 1990, ASTM: Phil., PA. p. 193-202.
19. Eyring, H. and E.M. Eyring, *Modern Chemical Kinetics*. Selected Topics in Modern Chemistry, ed. H.H. Sisler and C.A.V. Werf. 1963, New York: Reinhold Publ. Corp.
20. Uhlig, H.H., *Corrosion and Corrosion Control*. 2 ed. 1971, New York: J. Wiley & Sons.
21. Fontana, M.G. and N.D. Greene, *Corrosion Engineering*. McGraw-Hill Series in Materials Science and Engineering, ed. M.B. Bever, M.E. Shank, and C.A. Wert. 1978, New York: McGraw-Hill Book Co. 263.
22. Bates, D.M. and D.G. Watts, *Nonlinear Regression Analysis and Its Applications*. 1988, New York: J. Wiley & Sons, Inc.

Table A1 - Sample Burial Site Information

Site No.	Soil Type	Location	Internal Drainage	Burial Depth m	Sand %	Silt %	Clay %	Colloid %	Suspension %
1	Allis silt loam	Cleveland, OH	Poor	2.41	9.8	37.7	52.5	31.1	38.0
2	Bell clay	Dallas, TX	Poor	1.02	17.2	37.9	44.9	40.4	19.1
3	Cecil clay loam	Atlanta, GA	Good	0.76	29.0	24.9	46.1	37.8	4.4
4	Chester loam	Jenkintown, PA	Fair	0.91	29.3	53.0	17.7	9.9	27.9
5	Dublin clay adobe	Oakland, CA	Poor	0.76	25.6	38.6	35.8	30.2	40.8
6	Everett gravelly sandy loam	Seatle, WA	Good	0.91	69.0	23.8	7.2	3.5	12.2
7	Maddox silt loam	Cincinnati, OH	Fair	0.56	10.3	26.1	63.6	55.0	41.2
8	Fargo clay loam	Fargo, ND	Poor	1.68	2.2	27.7	70.1	50.7	20.2
9	Genesee silt loam	Sidney, OH	Poor	0.56					
10	Gloucester sandy loam	Middleboro, MA	Fair	0.91	64.0	29.4	6.6	2.8	16.4
11	Hagerstown loam	Baltimore, MD	Good	0.91	25.8	21.1	53.1	45.9	7.9
12	Hanford ne sandy loam	Los Angeles, CA	Fair	0.61					
13	Hanford very ne sandy loam	Bakers eld, CA	Fair	0.76					
14	Hempstead silt loam	St. Paul, MN	Fair	1.12	56.6	29.5	13.9	9.5	19.8
15	Houston black clay	San Antonio, TX	Poor	0.91	4.4	25.2	70.4	62.0	28.3
16	Kalmia ne sandy loam	Mobile, AL	Fair	0.76	50.4	23.1	26.5	21.8	11.1
17	Keyport loam	Alexandria, VA	Poor	0.91	9.6	38.6	51.7	39.6	43.5
18	Knox silt loam	Omaha, NB	Good	1.22	1.3	78.4	20.3	15.0	24.2
19	Lindley silt loam	Des Moines, IA	Good	0.91	15.7	50.1	34.2	29.3	16.9
20	Mahoning silt loam	Cleveland, OH	Poor	1.22	19.6	44.1	36.3	23.7	37.7
21	Marshall silt loam	Kansas City, MO	Fair	1.52	3.2	65.9	30.9	27.1	26.0
22	Memphis silt loam	Memphis, TN	Good	0.84	1.2	76.5	22.3	18.3	38.7
23	Merced silt loam	Buttonwillow, CA	Fair	0.76					
24	Merrimac gravelly sandy loam Norwood,	MA	Good	0.84	72.0	22.4	5.6	2.7	10.4
25	Miami clay loam	Milwaukee, WI	Fair	0.91	21.0	43.0	36.0	21.8	24.8
26	Miami silt loam	Spring eld, OH	Good	0.91					
27	Miller clay	Bunkie, LA	Poor	0.76	1.4	10.8	87.8	71.5	30.2
28	Montezuma clay adobe	San Diego, CA	Poor	1.02					
29	Muck	New Orleans, LA	Very Poor	0.61					
30	Muscatine silt loam	Davenport, IA	Poor	0.91	2.1	65.5	32.4	26.1	25.7
31	Norfolk ne sand	Jacksonville, FL	Good	0.61	97.3	2.1	0.6		1.8
32	Ontario loam	Rochester, NY	Good	1.22	42.1	42.1	15.8	8.2	15.9
33	Peat	Milwaukee, WI	Very Poor	0.61					
34	Penn silt loam	Norristown, PA	Fair	0.91					
35	Romona loam	Los Angeles, CA	Good	0.91	35.9	37.3	26.0	19.3	22.2
36	Ruston sandy loam	Meridian, MS	Good	0.91	60.6	21.8	17.6	14.8	17.0
37	St. John's ne sand	Jacksonville, FL	Poor	0.76	90.6	4.9	4.5	4.3	3.3
38	Sassafras gravelly sandy loam	Camden, NJ	Good	0.76					
39	Sassafras silt loam	Wilmington, DE	Fair	0.76	42.1	42.6	15.3	8.7	18.9
40	Sharkey clay	New Orleans, LA	Poor	0.76	2.5	50.4	47.1	32.8	24.9
41	Summit silt loam	Kansas City, MO	Fair	0.91	3.0	56.7	40.3	35.0	24.7
42	Susquehanna clay	Meridian, MS	Poor	0.76	30.1	24.1	45.8	40.9	11.8
43	Tidal marsh	Elizabeth, NJ	Very Poor	0.91					
44	Wabash silt loam	Omaha, NB	Good	0.76	2.4	66.4	31.2	25.8	22.1
45	Unidenti ed alkali soil	Casper, WY	Poor	0.76					
46	Unidenti ed sandy loam	Denver, CO	Good	1.27					
47	Unidenti ed silt loam	Salt Lake City, UT	Poor	0.91	9.0	44.9	46.1	27.7	45.5

Table A2 - Site characterization variables, corrosion damage measures, original units, SI units, conversion factors, and distribution type

Variable No.	Variable Name	Original Units	SI units	Conversion	Distribution Type	Note
1	Site No.	NA	NA	NA	NA	Arbitrary
2	Soil	NA	NA	NA	NA	Named after first location and particles size distr.
3	Location	NA	NA	NA	NA	Actual location
4	Internal Drainage	NA	NA	NA	NA	Arbitrary ranking based on site location, topography, and horizons.
5	Burial Depth	ft	m	0.3048 ft/m	Normal	Standard depth for burial location
6	Percent Sand in Soil	%	%	NA	NA	Particles 0.05-1.0 mm dia
7	Percent Silt in Soil	%	%	NA	NA	Particles 0.002-0.5 mm dia
8	Percent Clay in Soil	%	%	NA	NA	Particles <0.002 mm dia
9	Percent Colloid in Soil	%	%	NA	NA	
10	Percent Suspension	%	%	NA	NA	
11	Resistivity-Conductivity	ohm-cm	S/m	C=1/(R*0.01)	Log-Normal	Measured at 60 °F (15.6 °C) Conductivity prefered for analysis
12	Temperature	°F	°C	C=(F-32)*(5/9)	Normal	
13	Annual Precipitation	in/yr	mm/yr	25.4 mm/in	Normal	Estimated from nearest location of measurements
14	Moisture Equivalent	%	%	1	Log-Normal	
15	Air Pore Space	%	%	1	Log-Normal	
16	Density (Specific Gravity)	g/cm^3	kg/m^3	1000	Normal	Specific gravity units are that of the density of water.
17	Volume Shrinkage	%	%	1	Log-Normal	
18	pH	-log[mol/L]	-log[mol/L]	1	Normal	Hydrogen ion concnetration
19	Total Acidity	mg-eq/100 g	mol/kg	0.01	Log-Normal	Measure of acid buffering
20	[Na+K]	mg-eq/100 g	mol/kg	0.01	Log-Normal	Soluble ions per unit mass of soil - [Na] and [K] expressed as Na
21	[Ca]	mg-eq/100 g	mol/kg	0.01	Log-Normal	Soluble ions per unit mass of soil
22	[Mg]m	g-eq/100 g	mol/kg	0.01	Log-Normal	Soluble ions per unit mass of soil
23	[CO$_3$]m	g-eq/100 g	mol/kg	0.01	Log-Normal	Soluble ions per unit mass of soil
24	[HCO$_3$]m	g-eq/100 g	mol/kg	0.01	Log-Normal	Soluble ions per unit mass of soil
25	[Cl]	mg-eq/100 g	mol/kg	0.01	Log-Normal	Soluble ions per unit mass of soil
26	[SO$_4$]	mg-eq/100 g	mol/kg	0.01	Log-Normal	Soluble ions per unit mass of soil
A'	Corrosion Mass Loss (ML)	oz/ft^2	g/m2	2.634	NA	Measured after corrosion products removed; exposure times varied; average of 2 samples.
A	Corrosion Mass Loss Rate (MLR)	NA	g/m2/d	A'/exp time	Log-Normal	Assumes approximately linear (constant rate) behavior over exposure period
B'	Corrosion Penetraton (CP)	mils	mm or μm	25.4 μm/mil	NA	Measured maximum penetration; exposure times varied; average of 2 samples.
B	Corrosion Penetraton Rate (CPR)	NA	mm/yr	B'/exp time	Log-Normal	Assumes approximately linear (constant rate) behavior over exposure period
C	Pitting Ratio (PR)	NA	m/m	CPR/(MLR*K) K=0.04647	Log-Normal	Rato of maximum penetration to mean calculated from mass loss and density

Table A3 - Physical characteristics of soils at sites

Site No.	Resistivity, ohm-m	Mean Temperature °C	Annual Precipitation mm/yr	Moisture Equivalent %	Air-pore Space %	Specific Gravity kg/m3	Volume Shrinkage %
1	12.2	9.6	859	28.6	1.1		6.6
2	6.8	18.6	919	37.6	2	1950	23
3	300.0	16.2	1227	29.1	18.2	1600	7
4	66.7	12.2	1016	22.2	7	1780	2.2
5	13.5	13.6	594	28.8	4.9	2000	22.6
6	451.0	10.6	864	12.2	40.6	1500	0.1
7	21.2	11.8	980	34.3	3.7	2020	34.5
8	3.5	3.9	533	37	8.7	1560	21
9	28.2	10.7	991	24.8	15.8	1740	5.6
10	74.6	10.0	1041	13	27.8	1580	0.2
11	110.0	13.0	1082	32	15.5	1490	8.6
12	31.9	16.9	386	12.4	33.5		0
13	2.9	18.1	142	21.7	34.5		0
14	35.2	6.8	691	17.2	14.4	1760	1
15	4.9	20.5	691	51.4	5.7	2080	39.8
16	82.9	19.6	1565	22.2	12	1650	0.6
17	59.8	12.8	1067	30.8	4.4	1720	5.4
18	14.1	10.3	706	28.4	16.6	1260	1.3
19	19.7	9.7	813	28.4	3.9	1760	11.8
20	28.7	9.6	859	22.4	3.8	1900	3.9
21	23.7	12.4	942	31.2	10.8	1660	6.5
22	51.5	16.4	1212	28.4	9.6	1670	3
23	2.8	18.3	152	24.7	6.1	1690	0.2
24	114.0	10.0	1041	13	34.7	1400	0
25	17.8	7.8	765	25.8	9.5	1950	7.6
26	29.8	11.7	940	16.4	20.9	1950	1
27	5.7	19.4	1422	42.6	1.9	2010	32.5
28	4.1	16.1	262	24.6	2.5		5.9
29	12.7	20.7	1458	34.5	26.6		5.8
30	13.0	9.9	815	29.4	7.2	1810	7.5
31	205.0	20.7	1204	2.8	38.1	1550	0
32	57.0	8.7	833	17.8	11.7	1850	0.1
33	8.0	7.8	765	72.8	34		16.9
34	49.0	12.2	1016	23.4	11.7	1820	8.4
35	20.6	16.9	386	18	10.9	1890	3.1
36	112.0	17.8	1346	13.8	16	1620	0
37	112.0	20.7	1204	7			0
38	386.0	12.2	1016	3	32.1	1590	0
39	74.4	12.2	1016	24.2	7.5	1720	3.8
40	9.7	20.7	1458	33	2.3	1780	16.4
41	13.2	12.4	942	33.1	6.9	1610	14.6
42	137.0	17.8	1346	34.8	14.9	1790	4.7
43	0.6	11.1	1092	55.4			
44	10.0	10.3	706	31.2	7.2	1550	6
45	2.6	8.4	389	14.8	18.7		0
46	15.0	10.0	358	7.6	23.2		0
47	17.7	10.9	409	25.7	2.6	1720	3.7

Table A4 - Chemical characteristics of soils at sites

Site No.	pH	Total Acidity, mol/kg	Na+K as [Na] mol/kg	[Ca] mol/kg	[Mg] mol/kg	[CO_3] mol/kg	[HCO_3] mol/kg	[Cl] mol/kg	[SO_4] mol/kg
1	7.0	1.1E-01	7.2E-03	2.5E-03	4.3E-03	0.0E+00	9.0E-04	9.0E-04	8.3E-03
2	7.3	3.5E-02	2.8E-03	1.1E-02	1.3E-03	0.0E+00	1.2E-02	4.0E-04	1.8E-03
3	5.2	1.2E-01							
4	5.6	7.6E-02							
5	7.0	6.5E-02	9.3E-03	4.8E-03		0.0E+00	6.9E-03	3.0E-04	2.5E-03
6	5.9	1.3E-01							0.0E+00
7	4.4	3.0E-01							0.0E+00
8	7.6		1.4E-02	1.7E-02	2.6E-02	0.0E+00	7.1E-03	1.0E-04	4.4E-02
9	6.8	7.2E-02							
10	6.6	3.6E-02							
11	5.3	1.1E-01						0.0E+00	0.0E+00
12	7.1	2.5E-02	3.9E-03	5.0E-03	1.6E-03	0.0E+00	4.0E-03	0.0E+00	1.4E-03
13	9.5		6.2E-02	9.0E-04	1.3E-03	0.0E+00	1.1E-02	1.6E-02	3.8E-02
14	6.2	5.6E-02							
15	7.5	5.0E-02	2.2E-02	8.8E-03	2.0E-03	0.0E+00	2.0E-02	1.3E-03	7.3E-03
16	4.4	1.2E-01							
17	4.5	1.9E-01							
18	7.3	1.4E-02	2.7E-03	6.3E-03	2.0E-03	0.0E+00	9.4E-03	0.0E+00	2.5E-03
19	4.6	1.1E-01	3.8E-03	3.2E-03	4.1E-03	0.0E+00	1.6E-03	3.0E-04	4.6E-03
20	7.5	1.5E-02	2.5E-03	4.8E-03	2.0E-03	0.0E+00	5.1E-03	0.0E+00	1.5E-03
21	6.2	9.5E-02							
22	4.9	9.7E-02							
23	9.4		8.4E-02	3.8E-03	2.2E-03	2.0E-04	1.9E-02	1.1E-02	5.6E-02
24	4.5	1.3E-01							
25	7.2	4.7E-02	2.3E-03	7.0E-03	4.1E-03	0.0E+00	1.0E-02	3.0E-04	1.0E-03
26	7.3	2.6E-02	2.7E-03	5.0E-03	3.1E-03	0.0E+00	7.0E-03	3.0E-04	1.2E-03
27	6.6	3.7E-02	5.3E-03	1.9E-02	1.1E-02	0.0E+00	2.0E-02	8.0E-04	1.5E-02
28	6.8		1.5E-02	6.0E-04	1.8E-03	0.0E+00	1.2E-03	9.9E-03	8.9E-03
29	4.2	2.8E-01	2.2E-02	1.9E-02	1.6E-02	0.0E+00	0.0E+00	1.7E-02	2.3E-02
30	7.0	2.6E-02	3.2E-03	6.5E-03	4.0E-03	0.0E+00	7.1E-03	9.0E-04	2.4E-03
31	4.7	1.8E-02							
32	7.3	5.0E-03	2.3E-03	7.0E-03	1.2E-03	0.0E+00	7.3E-03	1.0E-04	4.2E-03
33	6.8	3.6E-01	1.5E-02	7.3E-02	4.1E-02	0.0E+00		2.3E-02	2.1E-02
34	6.7	7.0E-02							
35	7.3	5.7E-02	6.8E-03	6.8E-03	4.9E-03	0.0E+00	1.1E-02	6.0E-04	3.5E-03
36	4.5	4.6E-02							
37	3.8	1.5E-01							
38	4.5	1.7E-02							
39	5.6	6.6E-02							
40	6.0	9.4E-02	5.6E-03	5.8E-03	4.4E-03	0.0E+00	9.3E-03	7.0E-04	2.8E-03
41	5.5	1.1E-01	3.0E-03	5.4E-03	3.6E-03	0.0E+00	7.8E-03	4.0E-04	4.6E-03
42	4.7	2.8E-01							
43	3.1	3.7E-01	4.5E-01	5.2E-02	9.5E-02	0.0E+00	0.0E+00	4.3E-01	3.7E-01
44	5.8	8.8E-02	1.1E-02	1.1E-02	6.6E-03	0.0E+00	2.0E-02	8.2E-03	4.1E-03
45	7.4		8.2E-02	3.7E-02	7.0E-03	0.0E+00	2.4E-03	1.8E-03	1.2E-01
46	7.0								
47	7.6	3.0E-02	6.7E-03	7.2E-03	3.9E-03	0.0E+00	8.8E-03	6.0E-04	4.8E-03

Table A5 - Distribution and standard scores for physical characteristics of exposure sites.

	Drain	Cond	Temp	Ann Precip	Moist Equiv	Pore Spc	Density	Vol Shrink	Burial Depth
Dist.	NA	lognrm	norm	norm	lognrm	lognrm	norm	lognrm	norm
Mean	NA	-1.39	13.37	883.43	1.35	1.00	1729.19	0.63	0.92
Std Dev	NA	0.62	4.41	350.14	0.28	0.40	185.09	0.68	0.32
Units			°C	mm	%	%	kg/m3	%	m

Site No.		Standardized Scores (Z) for Site Characteristics							
		Z (1/R)	Z (T)	Z (AP)	Z (ME)	Z (PS)	Z (rho)	Z (VS)	Z (BD)
1	2	-0.70	-0.87	-0.07	0.39	-2.38		0.27	4.71
2	2	-1.05	1.19	0.10	0.82	-1.74	1.19	1.07	0.29
3	4	1.52	0.65	0.98	0.42	0.65	-0.70	0.31	-0.51
4	3	0.49	-0.26	0.38	-0.01	-0.38	0.27	-0.43	-0.03
5	2	-0.48	0.04	-0.83	0.40	-0.77	1.46	1.06	-0.51
6	4	2.60	-0.64	-0.06	-0.95	1.52	-1.24	-2.41	-0.03
7	3	0.10	-0.36	0.28	0.68	-1.07	1.57	1.33	-1.16
8	2	-1.20	-2.15	-1.00	0.80	-0.15	-0.91	1.02	2.38
9	2	0.15	-0.61	0.31	0.17	0.50	0.06	0.17	-1.16
10	3	0.65	-0.76	0.45	-0.85	1.11	-0.81	-1.97	-0.03
11	4	0.89	-0.08	0.57	0.57	0.48	-1.29	0.44	-0.03
12	3	0.34	0.80	-1.42	-0.93	1.31			-1.00
13	3	-1.48	1.08	-2.12	-0.04	1.34			-0.51
14	3	0.39	-1.50	-0.55	-0.41	0.40	0.17	-0.94	0.61
15	2	-1.06	1.62	-0.55	1.31	-0.60	1.90	1.43	-0.03
16	3	0.78	1.42	1.95	-0.01	0.20	-0.43	-1.26	-0.51
17	2	0.46	-0.13	0.52	0.51	-0.88	-0.05	0.14	-0.03
18	4	-0.25	-0.69	-0.51	0.38	0.55	-2.53	-0.77	0.93
19	4	-0.10	-0.83	-0.20	0.38	-1.01	0.17	0.65	-0.03
20	2	0.22	-0.87	-0.07	0.01	-1.04	0.92	-0.06	0.93
21	3	0.12	-0.21	0.17	0.53	0.09	-0.37	0.26	1.90
22	4	0.43	0.70	0.94	0.38	-0.04	-0.32	-0.23	-0.27
23	3	-1.75	1.13	-2.09	0.16	-0.53	-0.21	-1.97	-0.51
24	4	1.25	-0.76	0.45	-0.85	1.35	-1.78		-0.27
25	3	-0.11	-1.26	-0.34	0.23	-0.05	1.19	0.36	-0.03
26	4	0.23	-0.39	0.16	-0.49	0.80	1.19	-0.94	-0.03
27	2	-1.06	1.38	1.54	1.02	-1.79	1.52	1.30	-0.51
28	2	-1.07	0.62	-1.78	0.15	-1.49		0.20	0.29
29	1	-0.62	1.67	1.64	0.69	1.06		0.19	-1.00
30	2	-0.59	-0.78	-0.19	0.43	-0.35	0.44	0.36	-0.03
31	4	1.49	1.67	0.92	-3.27	1.45	-0.97		-1.00
32	4	0.44	-1.07	-0.14	-0.36	0.17	0.65	-2.41	0.93
33	1	-0.85	-1.26	-0.34	1.86	1.33		0.88	-1.00
34	3	0.42	-0.26	0.38	0.07	0.17	0.49	0.43	-0.03
35	4	0.02	0.80	-1.42	-0.34	0.10	0.87	-0.21	-0.03
36	4	1.02	1.00	1.32	-0.76	0.51	-0.59		-0.03
37	2	1.13	1.67	0.92	-1.83				-0.51
38	4	1.56	-0.26	0.38	-3.16	1.26	-0.75		-0.51
39	3	0.63	-0.26	0.38	0.13	-0.31	-0.05	-0.08	-0.51
40	2	-0.78	1.67	1.64	0.62	-1.58	0.27	0.86	-0.51
41	3	-0.52	-0.21	0.17	0.62	-0.40	-0.64	0.78	-0.03
42	2	1.36	1.00	1.32	0.70	0.43	0.33	0.06	-0.51
43	1	-2.04	-0.51	0.60	1.43				-0.03
44	4	-0.78	-0.69	-0.51	0.53	-0.35	-0.97	0.21	-0.51
45	2	-1.93	-1.12	-1.41	-0.65	0.68			-0.51
46	4	-0.18	-0.76	-1.50	-1.70	0.91			1.09
47	2	-0.14	-0.56	-1.36	0.22	-1.45	-0.05	-0.10	-0.03

Table A6 - Distribution and standard scores for chemical characteristics of exposure sites.

	pH	Tot Acid	[Na+K]	[Ca]	[Mg]	[HCO3]	[Cl]	[SO4]
Dist.	norm	logn	logn	logn	logn	logn	logn	logn
Mean	6.16	-1.17	-2.04	-2.14	-2.35	-2.16	-2.93	-2.14
Std Dev	1.41	0.42	0.58	0.46	0.48	0.37	0.73	0.66
Units	mol/L	mol/kg	mol/kg	mol/kg	mol/kg	mol/kg	mol/kg	mol/kg

	Standardized Scores (Z) for Site Characteristics							
Site No.	Z (pH)	Z (TA)	Z [Na]	Z [Ca]	Z [Mg]	Z [HCO3]	Z [Cl]	Z [SO4]
1	0.59	0.55	-0.18	-1.00	-0.04	-2.36	-0.15	0.09
2	0.81	-0.68	-0.88	0.38	-1.12	0.62	-0.63	-0.92
3	-0.68	0.56						
4	-0.40	0.12						
5	0.59	-0.04	0.01	-0.39		0.00	-0.80	-0.70
6	-0.19	0.67						
7	-1.25	1.55						
8	1.02		0.32	0.81	1.57	0.03	-1.45	1.20
9	0.45	0.07						
10	0.31	-0.65						
11	-0.61	0.49						
12	0.66	-1.04	-0.64	-0.35	-0.94	-0.63		-1.08
13	2.37		1.42	-1.96	-1.12	0.56	1.57	1.09
14	0.03	-0.19						
15	0.95	-0.31	0.64	0.18	-0.73	1.23	0.07	0.01
16	-1.25	0.58						
17	-1.18	1.09						
18	0.81	-1.64	-0.91	-0.14	-0.73	0.36		-0.70
19	-1.11	0.50	-0.66	-0.77	-0.09	-1.69	-0.80	-0.30
20	0.95	-1.57	-0.97	-0.39	-0.73	-0.35		-1.04
21	0.03	0.36						
22	-0.90	0.38						
23	2.30		1.64	-0.61	-0.65	1.16	1.34	1.35
24	-1.18	0.65						
25	0.74	-0.38	-1.03	-0.04	-0.09	0.44	-0.80	-1.31
26	0.81	-0.99	-0.91	-0.35	-0.34	0.02	-0.80	-1.19
27	0.31	-0.63	-0.41	0.88	0.82	1.23	-0.22	0.50
28	0.45		0.37	-2.34	-0.83	-2.03	1.27	0.14
29	-1.39	1.49	0.63	0.91	1.12		1.59	0.77
30	0.59	-0.99	-0.78	-0.11	-0.11	0.03	-0.15	-0.73
31	-1.04	-1.38						
32	0.81	-2.71	-1.03	-0.04	-1.20	0.07	-1.45	-0.36
33	0.45	1.75	0.38	2.16	1.99		1.76	0.72
34	0.38	0.04						
35	0.81	-0.18	-0.22	-0.06	0.08	0.54	-0.39	-0.48
36	-1.18	-0.40						
37	-1.68	0.85						
38	-1.18	-1.44						
39	-0.40	-0.02						
40	-0.12	0.35	-0.37	-0.21	-0.02	0.35	-0.30	-0.63
41	-0.47	0.51	-0.83	-0.28	-0.20	0.14	-0.63	-0.30
42	-1.04	1.49						
43	-2.17	1.77	2.89	1.84	2.75			2.61
44	-0.26	0.28	0.10	0.37	0.35	1.22	1.16	-0.37
45	0.88		1.62	1.53	0.40	-1.22	0.26	1.86
46	0.59							
47	1.02	-0.84	-0.23	-0.01	-0.13	0.28	-0.39	-0.27

Table A7 - Site soil horizons

Site No.	Burial Depth, m	Horizon	Depth Range in		Depth Range m		Horizon Description	Int. Drainage	Topography
1	2.41	1	0	8	0.00	0.20	grayish yellow or yellowish gray silt loam mottled with yellow and yellowish brown	Poor	Undulating to gently rolling
		2	8	23	0.20	0.58	mottled yellow and gray silty clay loam which contains fragments of shale		
		3	23	30	0.58	0.76	bluish gray silty clay loam with bands of yellow indicating the bedding planes of the shale.		
		4	30	70	0.76	1.78	silty clay or silty clay loam layer of shale which has a bluish gray color and is streaked along bedding planes with		
		5	70	76	1.78	1.93	reddish brown shale streaked with gray		
		6	76	90	1.93	2.29	compact bluish gray shale with yellowish brown and reddish brown streaks		
		7	90	100	2.29	2.54	the streaks become less conspicuous. This shale runs high in aluminum sulfate, which, with water, breaks down into aluminum hydroxide and sulfuric acid.		
2	1.02	1	0	10	0.00	0.25	black to dark brown silty clay	Poor	Level
		2	10	740	0.25	18.80	black clay. No definitely residual matter was discovered within 40 inches. Small rounded quartzite gravel and lime concentrations disseminated through the subsoil		
3	0.76	1	0	8	0.00	0.20	grayish brown, rather compact, very fine sandy loam. A few fragments of granite and quartz found on the surface.	Little Excessive	Moderate slope
		2	8	10	0.20	0.25	transition layer into…		
		3	10	32	0.25	0.81	compact brittle red clay containing very few mica flakes and practically no sand and stones		
		4	32	48	0.81	1.22	micaceous, more friable, and not as compact as above horizon, red clay loam or clay.		
		5	48	52	1.22	1.32	layer of sandy clay with yellowish mottlings		
		6	52	70	1.32	1.78	red micaceous clay as in 32-48		
		7	70	74	1.78	1.88	red very fine sandy loam with yellowish mottlings		
		8	74	90	1.88	2.29	Moderately friable red very fine sandy loam, full of mica crystals, and having a few brownish and yellowish mottlings due to partially decomposed rock.		
		9	96	108	2.44	2.74	very friable fine sandy loam, mottled yellow, red, and		
4	0.91	1	0	6	0.00	0.15	The top 6 inches of the trench is a mixture of road material and soil. No vegetation.	Good	Gently rolling
		2	0	10	0.00	0.25	grayish brown mellow loam gradually getting lighter in color with increasing depth.		
		3	10	34	0.25	0.86	mellow, only slightly darker in color and heavier in texture with increasing depth.		
		4	34	96	0.86	2.44	micaeous rather loose friable silt loam containing considerable fine sand. At 36 inches there is a layer of partially decomposed granite.		
5	0.76	1	1	10	0.03	0.25	dark dull grey or drab clay of adobe structure, sticky when wet, contains numerous plant and grass roots and a appreciable amount of fine gritty material and gravel	Poor	Smooth and level
		2	10	36	0.25	0.91	slightly more compact brownish gray or drab friable clay which is sticky when wet. Somewhat mottled with brown and dull slaty gray or black streaks. It contains spherical shotlike iron concretions of black or bluish black color, ranging in size from a pinhead to small buckshot.		
		3	36	48	0.91	1.22	soil grades into a yellowish brown slity clay material. This horizon is mildly calcerous and is the upper limit of lime		
		4	48	60	1.22	1.52	yellowish brown compact clay containing many light grayish fragments of lime carbonate nodules localized in thin seams or layers, the material being partially cemented.		

Table A7 - Site soil horizons

Site No.	Burial Depth, m	Horizon	Depth Range in		Depth Range m		Horizon Description	Int. Drainage	Topography
6	0.91	1	0	8	0.00	0.20	brown to light brown sandy loam darkened by presence of organic matter.	Excessive	Moderately rolling
		2	8	24	0.20	0.61	light brown sandy loam; Both this and the above horizon contain little gravel and considerable coarse sand. Both horizons are loose and friable and contain numerous grass		
		3	24	30	0.61	0.76	grayish brown gravelly sandy loam. Slightly compact. Below 30 inches hard and cemented gravel and sand, with very little lime of a grayish brown color.		
7	0.56	1	0	5	0.00	0.13	brownish yellow friable silty clay loam	Fair	Smooth ridge top
		2	5	15	0.13	0.38	brownish yellow smooth, plastic, heavy, moderately tight clay mottled light grey. The mottles are of moderate extent and development and occur in small irregular veins. The soil material fractures into irregularly shaped lumps, ranging in size from 0.5 to 1.5 inches in diameter.		
		3	15	22	0.38	0.56	brownish yellow or yellowish sticky, plastic, slowly pervious moderately compact heavier clay containing a moderate amount of light grey mottles. It has a fragmental structure forming hard, irregularaggregates from 0.5 to 1.5 inches in		
		4	22	30	0.56	0.76	varicolored bluish grey and olive-green tight smooth, plastic, very heavy clay or silty clay having occasional staining of rust yellow. This layer has been developed from the weathering of underlying shale rock materials.		
8	1.68	1	0	24	0.00	0.61	black noncalcareous clay loam. Rather friable. Breaks with concoidal fracture into pea-size pieces.	Poor	Level
		2	24	42	0.61	1.07	calcareous transition layer with tongues of both horizons extending into the layer.		
		3	42	88	1.07	2.24	grayish brown heavy clay loam. Light gray when dry-highly calcareous.		
		4	88	+	2.24	+	parent material of old lake laid deposits; grayish brown color containing rusty brown streaks and mottlings; few hard concretions that are largely lime.		
9	0.56	1	0	10	0.00	0.25	brownish gray silt loam, slightly streaked with reddish	Poor	
		2	10	16	0.25	0.41	gray loam streaked reddish brown and mottled yellowish brown and brownish yellow.		
		3	16	22	0.41	0.56	transition to fine sandy loam mottled reddish brown. At 22 bed of gray gravel		
10	0.91	1					Surface-light brown sandy loam.	Fair	
		2					Subsoil-light grayish brown fine sandy loam containing some gravel.		
11	0.91	1	0	12	0.00	0.30	dark brown or brown friable loam.	Good	Slight Slope
		2	12	33	0.30	0.84	reddish brown or red clay loam. Moderately compact. Contains fragments of stone, chert.		
		3	33	+	0.84	+	moderately friable rusty brown heavy silt loam with a reddish cast. This extends to the underlying rock, which is rather clear, crystalline, and hard (not limestone). In one place in the trench the rock is at a depth of about 4		
12	0.61						The entire profile is a grayish brown friable, loose, micaceous fine sandy loam containing thin layers of material as heavy as loam and as tight as sand. Noncalcareous at surface, and only faintly calcareous at 6 This soil differs from soil 13 in that it does not contain soluble carbonates in appreciable amount.	Good	Practically Level

Table A7 - Site soil horizons

Site No.	Burial Depth, m	Horizon	Depth Range in		Depth Range m		Horizon Description	Int. Drainage	Topography
13a	0.76	1	0	56	0.00	1.42	light grayish brown smooth, friable, micaceous very fine sandy loam.	Fair	Almost Level
		2	56	62	1.42	1.57	light grayish brown very fine sand.		
		3	62	68	1.57	1.73	same as 0-56.		
		4	68	72	1.73	1.83	same as 56-62.		
							The soil is high in alkali in the carbonate form, and formerly called black alkali.		
13b		1	0	6	0.00	0.15	grayish brown very slightly compacted loam.	Good	Very Gentle Undulating
		2	6	84	0.15	2.13	light grayish brown friable loose micaceous very fine sandy loam. Numerous roots in first 3 feet. Few light colored		
							A special set of specimens are buried at the site. The profile is similar to site 13a, but differs by being low in alkali		
14	1.12	1	0	15	0.00	0.38	dark brown (almost black) silt loam.	Fair	Very Gentle Undulating
		2	15	24	0.38	0.61	transition layer consisting of tongues and streaks of the two adjoining horizons extending into each other.		
		3	24	42	0.61	1.07	brown silt loam with yellowish cast, slightly compact.		
		4	42	+	1.07	+	grayish brown sand containing some gravel.		
							Entire profile is noncalcareous.		
15	0.91	1	0	36	0.00	0.91	black clay with no appreciable change. Highly calcareous. Small fragments of lime are found throughout the section.	Poor	
16	0.76	1	0	8	0.00	0.20	grayish brown fine sandy loam, which appears to have been disturbed.	Fair	Gentle Slope
		2	8	42	0.20	1.07	yellowish brown very fine sandy loam. Texture gradually gets finer and compactness increases with depth, home reddish mottlings and a few iron concretions about ^ inch in diameter, which are most numerous at about 3 feet and		
		3	42	48	1.07	1.22	brownish yellow or yellow silt loam mottled with red.		
		4	48	96	1.22	2.44	mottled red, gray, and yellow material containing thin layers of clay and fine sand but with the average texture of Below 72 inches the color is light yellowish brown with light gray mottlings.		
17	0.91	1	0	6	0.00	0.15	grayish brown loam or silt loam without structure. Moderately loose and friable.	Fair	Gentle Slope
		2	6	14	0.15	0.36	transition layer, slightly compact clay loam.		
		3	14	48	0.36	1.22	light yellowish brown rather compact clay loam with concoidal fracture exposing shiny surfaces. Slightly mottled with gray. Texture gets a little lighter with		
		4	48	74	1.22	1.88	brown fine sandy loam with slight reddish cast.		
		5	74	76	1.88	1.93	light gray clayey sand.		
		6	76	96	1.93	2.44	brown sand almost saturated with water.		
		7	96	+	2.44	+	gravel.		
							Entire profile is noncalcareous.		
18	1.22	1	0	8	0.00	0.20	dark brown silt loam full of brickbats, plaster, rotten wood, etc. The surface soil partly removed and mixed with foreign	Good	Practically Level
		2	8	72	0.20	1.83	light brown very uniform smooth friable silt loam that gets a little lighter in color with depth. Moderately moist. Contains a few brown spots due to rotten roots at 8 to 24 inches. Very faintly calcareous at 48 inches and below.		
19	0.91	1	0	4	0.00	0.10	dark brown silt loam, friaole and full of organic matter. .	Good	Moderate Slope
		2	4	18	0.10	0.46	slightly compact heavy silt loam, yellowish brown.		
		3	18	34	0.46	0.86	transition layer into. .		
		4	24	50	0.61	1.27	rather compact more yellowish brown clay containing a few dark-colored specks.		
		5	50	76	1.27	1.93	grayish brown clay loam with bright yellow mottlings and a few white specks. Less compact than above.		
		6	76	84	1.93	2.13	gritty material of variable texture and color, containing light colored cherty material.		
		7	84	2.13			large boulder or gravel.		

Table A7 - Site soil horizons

Site No.	Burial Depth, m	Horizon	Depth Range in		Depth Range m		Horizon Description	Int. Drainage	Topography
20	1.22	1	0	4	0.00	0.10	brownish gray heavy silt loam or light silty clay loam. .	Poor	Gently Undulating
		2	4	8	0.10	0.20	pinkish red clay, mottled brownish yellow, yellow, yellowish brown, and gray.		
		3	8	24	0.20	0.61	mottled drabbish gray-yellow, brownish yellowy and yellowish brown clay.		
		4	24	46	0.61	1.17	drabbish gray clay, mottled with brownish yellow, and		
		5	46	50	1.17	1.27	mottled gray, brownish yellow, and yellowish brown, calcareous clay.		
21	1.52	1	0	28	0.00	0.71	brown or chocolate brown friable, uniform silt loam. .	Good	Moderately Rolling
		2	28	36	0.71	0.91	transition layer.		
		3	36	84	0.91	2.13	light brown silt loam very uniform and smooth. Non-calcareous to 6 feet.		
		4	84	+	2.13	+	light brown noncalcareous clay slightly mottled with grayish brown.		
22	0.84	1	0	4	0.00	0.10	light brown silt loam containing thin discontinuous layers of darker color probably, due to the turning under of organic matter when the soil was cultivated.	Good	Very Gently Undulating
		2	4	96	0.10	2.44	light brown slightly compact silt loam with some grayish mottlings but no hard lime concretions. Very uniform in		
23	0.76	1	0	14	0.00	0.36	dark brown (almost black) silt loam. 1/4-inch crust, 3-inch mulch, which is underlaid by slightly compact very lightly moist material with no definite structure.	Fair	Level
		2	14	72	0.36	1.83	light gray loam, moderately compact and moist with some what lighter texture and a more open structure below 48 inches, where thin layers of sandy loam occur. Friable and Thin layers of grayish brown sand occur at 60 inches. Location has all indications of a soil high in alkali. Highly calcareous up to surface.		
24	0.84	1	0	4	0.00	0.10	brown loam containing considerable sand and coarse	Good	
		2	4	33	0.10	0.84	grayish coarse sand or fine gravel.		
25	0.91	1	0	6	0.00	0.15	grayish brown silt loam.	Fair	
		2	6	30	0.15	0.76	yellowish brown, stiff, heavy clay loam to clay, containing a small amount of gritty material.		
		3	30	48	0.76	1.22	slightly calcareous brownish yellow heavy clay loam, somewhat lighter than the above and also contains some		
26	0.91	1	0	2	0.00	0.05	grayish brown silt loam.	Good	
		2	2	7	0.05	0.18	brownish gray to yellowish-gray silt loam.		
		3	7	10	0.18	0.25	gray silt loam mottled faintly with yellow.		
		4	10	16	0.25	0.41	mottled yellow and gray silt loam.		
		5	16	24	0.41	0.61	brown clay loam to clay mottled brownish yellow and yellowish brown.		
		6	24	36	0.61	0.91	reddish brown stiff clay.		
		7	36	48	0.91	1.22	yellowish brown gravelly friable clay, somehwat calcareous in the lower part of the layer.		
27	0.76						Dull red heavy calcareous clay extending down below the depth at which the specimens are buried. Soil map shows Miller clay at this location and a sample of the soil was identified as typical Miller clay.	Very Poor	Level
28	1.02	1	0	8	0.00	0.20	filled material-brickbats, gravel, etc.	Poor	Level to Gently Rolling
		2	8	46	0.20	1.17	gray or light grayish-brown adobe containing some gritty material and gravel in the first foot. N oncalcareous.		
		3	46	50	1.17	1.27	light gray sandy clay, somewhat sticky.		
		4	50	60	1.27	1.52	grayish brown or yellowish brown gravelly sand.		
		5	60	+	1.52	+	gravel.		

Table A7 - Site soil horizons

Site No.	Burial Depth, m	Horizon	Depth Range in		Depth Range m		Horizon Description	Int. Drainage	Topography
29	0.61						Surface - to varying depths consists of dark colored material of variable texture, most of which is fill.	Very Poor	
							Subsoil - black, semifluid mass of well-decomposed mulch which rests upon an almost solid mat of old cypress stumps and roots that are in an excellent state of		
							Substratum - stiff, putty-like gray clay.		
							The land was originally a cypress swamp.		
30	0.91	1	0	6	0.00	0.15	dark brown silt loam (grayish brown when dry).	Poor	Level
		2	6	72	0.15	1.83	gray or grayish brown silt loam with yellow mottlings that are evenly distributed and containing a few brown specks. Non-calcareous throughout.		
31	0.61	1	0	4	0.00	0.10	grayish brown fine sand containing organic matter.	Good	Almost Level
		2	0	15	0.00	0.38	gradual transition into very slightly compact, very pale yellow sand. Deepest in color and more compact at 15		
		3	15	+	0.38	+	compactness gradually decreases and the color gets a little lighter. Slight yellow mottlings at 60 inches. The same sand probably extends to 20 or 30 feet. This soil was called Norfolk sand in previous soil corrosion reports.		
32	1.22	1	0	8	0.00	0.20	brown to grayish brown (when dry) mellow and friable, fine sandy loam to fine sand.	Good	Gently Sloping to Undulating
		2	8	18	0.20	0.46	slightly more compact, though crumbly loam to fine sandy loam, light brown to yellowish brown in color.		
		3	18	33	0.46	0.84	grayish brown to brownish gray compact loam in place, though friable when bored out.		
		4	33	+	0.84	+	partially weathered till material.		
							Parent material from which the soil is derived is largely limestone, with some sandstone, shale, and igneous rocks. Gravel and small stones are abundant in lower portions. The soil is cal-careous at from 15 to 24 inches.		
33	0.61						A black well-decomposed peat 30 to 36 inches deep, where it rests on a drab or bluish plastic clay loam. The lower part of the section was saturated with water. The peat merges into clyde loam, the line of separation being rather indefinite. A sample of this soil lost 42 percent on	Very Poor	
34	0.91	1	0	8	0.00	0.20	brown or dark brown silt loam.	Fair	Gentle Slope
		2	8	24	0.20	0.61	reddish brown silt loam containing considerable sand.		
		3	24	38	0.61	0.97	slightly lighter in color than above layer.		
		4	38	56	0.97	1.42	Indian red or reddish-brown silt loam.		
		5	56	+ 1.42		+	shale.		
35	0.91	1	0	22	0.00	0.56	light brown moderately compact loam w^ith slight reddish tint and a slight admixture of organic matter to 2 inches of surface. Very dry.	Good	Moderately Rolling
		2	22	54	0.56	1.37	slightly moist, hard, gritty, compact, brittle, reddish brown clay loam containing numerous white specks.		
		3	54	72	1.37	1.83	light reddish brown or light-brown gritty silt loam. White specks present but not as compact as horizon above. Entire profile is noncalcareous.		
36	0.91	1	0	8	0.00	0.20	light brown, loose, friable sandy loam. .	Good	Gently Rolling
		2	8	30	0.20	0.76	brownish red or rusty brown heavy fine sandy loam. Rather compact and hard. -		
		3	30	60	0.76	1.52	reddish brown, rather compact, heavy fine sandy loam. . . .		
		4	60	96	1.52	2.44	mottled red and yellow compact heavy fine sandy loam. . . No gravel or stones present in the profile.		

Table A7 - Site soil horizons

Site No.	Burial Depth, m	Horizon	Depth Range in		Depth Range m		Horizon Description	Int. Drainage	Topography
37	0.76	1	0	2	0.00	0.05	dark gray or grayish brown fine sand. The organic matter imparts the dark color.	Poor	Practically Level
		2	2	10	0.05	0.25	the material merges into a rather compact yellowish layer having a distinct lower boundary. The organic matter decreases with depth and the yellow color becomes brighter. The yellow sand contains a few very hard round black iron con-cretions about 1/4 inch in diameter that are		
		3	10	28	0.25	0.71	light gray slightly compact fine sand which becomes lighter with increasing depth and is almost white at 28 inches.		
		4	28	36	0.71	0.91	dark brown hard compact iron cemented hardpan with the characteristic coffee ground color.		
		5	36	60	0.91	1.52	pale yellow fine sand saturated with water.		
38	0.76	1	0	8	0.00	0.20	grayish brown gravelly sandy loam which gradually changes into a light yellowish brown or yellowish gray.	Good	Moderate Uniform Slope
		2	8	28	0.20	0.71	light gray or yellowish brown gravelly sandy loam which is darker than the horizon below.		
		3	28	96	0.71	2.44	light gray gravelly sandy loam with faint yellow cast. . Entire profile is loose and open and is noncalcareous. The amount of gravel is rather small for a gravelly type soil. The size.-of the gravel varies up to 8 inches in diameter and is all smooth and water-worn.		
39	0.76	1	0	12	0.00	0.30	This soil has been so disturbed that an accurate, description of the profile is impossible. grayish brown moderately friable silt loam.	Fair	Practically Level
		2	0	30+	0.00	+	slightly yellowish-brown silt loam which extends below the specimens. The trench bottom shows considerable gravel and a little gravel exists throughout the profile.		
40	0.76	1	0	8	0.00	0.20	dark brown or brown clay loam containing organic matter and full of grass roots. Rather compact.	Poor	Gently Undulating to Level
		2	8	30	0.20	0.76	stiff, plastic gray clay mottled with rusty colored material.		
		3					No definite hard iron concretions.		
		4	30	60	0.76	1.52	gray silt loam mottled with rusty brown. The rusty colored spots get lighter in color with depth and practically disappeared at 60 inches.		
41	0.91	1	0	22	0.00	0.56	very uniform and smooth brown silt loam.	Fair	Gentle Slope
		2	22	36	0.56	0.91	light brown smooth silt loam. .		
		3	36	108	0.91	2.74	light brown uniform silt loam faintly mottled with grayish Noncalcareous to 9 feet at which depth the soil is underlain by shale.		
42	0.76	1	0	6	0.00	0.15	Top soil corroded away. rather compact but friable light reddish brown clay.	Fair	Steep Slope
		2	6	45	0.15	1.14	mottled red, yellow, and gray very hard compact clay that has a cubical structure.		
		3	45	56	1.14	1.42	mottled red, yellow and gray heavy silt loam.		
		4	56	84	1.42	2.13	same as 6-45.		
43	0.91						Entire soil profile, and especially the surface foot, contains a large percentage of undecayed organic matter and has a black color when wet. Upon drying the color changes to grayish brown. The soil contains hydrogen sulfide and a considerable amount of soluble salts, but no lime. The surface portion of the soil lost 20.7 percent on ignition.	Very Poor	Level
44	0.76						Except for the addition of grass roots to the top 8 to 12 inches, the entire profile consists of a uniform dark brown silt loam (black when wet) or silty clay loam, to a depth of at least 8 feet. Non-calcareous throughout.	Good	Practically Level

Table A7 - Site soil horizons

Site No.	Burial Depth, m	Horizon	Depth Range in		Depth Range m		Horizon Description	Int. Drainage	Topography
45	0.76	1	0	6	0.00	0.15	light gray to light grayish brown sand to heavy silt loam. Little organic matter.	Poor	Level
		2	6	20	0.15	0.51	brown to grayish brown heavy, compact, gritty clay. Plastic and waxy when wet, but becomes hard and tough when		
		3	20	30	0.51	0.76	abrupt change to a light gray sandy clay. More friable than upper horizon due to higher sand content.		
		4	30	48	0.76	1.22	sand content decreases, color slightly darker and texture more compact than above horizon.		
							Type is highly alkaline, and white streaks and splotches of concentrated salts occur abundantly throughout the profile except in the surface soil.		
46	1.27	1	0	12	0.00	0.30	brown or light brown sandy loam.	Good	Very Gentle, Uniform Slope
		2	12	14	0.30	0.36	layer of brickbats and debris.		
		3	14	20	0.36	0.51	light brown sandy loam. All the above material is loose and		
		4	20	22	0.51	0.56	hard compact layer of cinders. All the above material is full and the next horizon is probably the original surface of the		
		5	22	36	0.56	0.91	hard, compact brown sandy loam.		
		6	36	120	0.91	3.05	light brown sandy loam which gets a little lighter in color and is calcareous below 60 inches, where it is slightly		
47	0.91	1	0	12	0.00	0.30	grayish brown or brown silt loam containing considerable organic matter. Highly calcareous at all depths.	Poor	Moderate Slope
		2	12	72	0.30	1.83	light gray moderately compact clay containing occasional mottlings of brownish yellow and reddish brown. A few lime concretions and occasional water-worn pebbles that are partly coated with lime are present.		

Table B1 - Sample Designations, Sizes, Alloys, and Compositions

Sample Alloy Desgn.	Material	Form	Nom.* Dia. (in)	Length (in)	Length (m)	Wall Thk. (in)	Wall Thk. (mm)	Alloy Composition, Mass Fraction in % (Balance Fe)					
								C	Si	Mn	S	P	Cu
a	Open-Hearth Iron Pipe	Pipe	1.5**	6.0	0.152	0.145	3.68	0.020	0.090	Tr†	0.050	0.010	0.014
b	Wrought Iron Pipe, Hand Puddled	Pipe	1.5**	6.0	0.152	0.145	3.68	0.030	0.150	Tr†	0.023	0.145	0.020
e	Bessemer Steel Pipe	Pipe	1.5**	6.0	0.152	0.145	3.68	0.090	–	0.390	0.040	0.088	–
y	Bessemer Steel Pipe	Pipe	1.5**	6.0	0.152	0.145	3.68	–	–	0.380	0.050	0.092	–
B	Wrought Iron Pipe, Hand Puddled	Pipe	3***	6.0	0.152	0.216	5.49	0.020	0.150	0.033	0.022	0.195	0.030
K	Open-hearth Steel Pipe	Pipe	3***	6.0	0.152	0.216	5.49	0.120	–	0.410	0.036	0.043	–
M	Bessemer Steel Pipe	Pipe	3***	6.0	0.152	0.216	5.49	0.080	–	0.400	0.038	0.098	–
Y	Copper-bearing Steel Pipe	Pipe	3***	6.0	0.152	0.216	5.49	0.070	–	0.240	0.032	0.008	0.220

* The nominal diameters and wall thicknesses for these samples correspond to ASME (ANSI) B36.10 Schedule 40 Pipe. Since the dimension in this schedule were adopted in the early 1900s from the iron pipe standards (IPS) adopted in the early 1800s, these samples almost certainly conformed to the nominal pipe sizes (NPS) of this schedule.

** NPS 1.5 inch pipe: External Diameter=1.900 in (48.26 mm), Wall Thickness=0.145 in (3.68 mm), Internal Diameter=1.610 in (40.89 mm), External Area per Unit Length= 0.497 ft (0.152 m)

*** NPS 3.0 inch pipe: External Diameter=3.500 in (88.90 mm), Wall Thickness=0.216 in (5.49 mm), Internal Diameter=3.068 in (78.93 mm), External Area per Unit Length= 0.916 ft (0.279 m)

† Trace: This usually means that the element was detected, but at too low a level to quantify. However, the detection limits or uncertainty levels of the eqipment were not specied.

Table B2 - Measured Sample Mass Loss (pg. 1 of 6)

Site No.	Exp Time yrs	Mass Loss (g/m^2) - 2 Sample Average Site								Average for Retrieval (g/m^2)				
		Nom. 1.5 in. Pipe				Nom. 3.0 in. Pipe				Mean	Std. Dev.	Min.	Max.	Range
		a	b	e	y	B	K	M	Y					
1	1.0	336	366	336	336	366	427	397	366	366	33	336	427	92
	3.6	1007	1221	1190	1190	1282	1282	1312	1160	1205	96	1007	1312	305
	5.5	1343	1678	1678	1434	1648	1648	1556	1404	1549	136	1343	1678	336
	7.7	1526	2075	2075	2014	2014	1709	2258	2014	1961	232	1526	2258	732
	9.6	2472	2868	2960	2685	2868	3082	2716	2777	2804	186	2472	3082	610
	11.6	2716	2685	2136	2838	2899	2899	3540	2685	2800	386	2136	3540	1404
2	2.1	580	610	549	732	580	671	671	549	618	67	549	732	183
	4.0	885	915	915	915	1068	824	946	763	904	89	763	1068	305
	5.9	1160	1282	1190	1038	1099	1221	1282	1282	1194	91	1038	1282	244
	7.9	1160	1129	1129	1099	1129	946	1190	702	1060	162	702	1190	488
	12.0	1831	2167	1770	1800	1922	1648	2014	1800	1869	161	1648	2167	519
	17.6	2380	2563	2380	2472	2472	2350	2167	2258	2380	126	2167	2563	397
3	2.0	458	549	519	610	488	610	641	519	549	65	458	641	183
	4.1	885	977	1007	1068	1099	1007	1099	1190	1041	93	885	1190	305
	6.0	977	1282	1129	1068	1160	1099	1099	1282	1137	104	977	1282	305
	8.0	1129	1343	1282	1190	1129	1251	1373	1160	1232	95	1129	1373	244
	10.1	1282	1404	1312	1251	977	1099	1160	1221	1213	133	977	1404	427
	12.1	1129	1495	1434	1556	1404	1282	1312	1373	1373	134	1129	1556	427
4	1.4	458	366	458	427	427	427	427	458	431	30	366	458	92
	4.0	1129	1068	1068	1068	1129	1129	1068	1129	1099	33	1068	1129	61
	6.1	1373	1343	1343	1404	1434	1404	1343	1495	1392	54	1343	1495	153
	8.0	1587	1648	1648	1617	1678	1678	1648	1770	1659	54	1587	1770	183
	12.0	2136	2014	2136	1892	2167	2136	2106	2289	2109	116	1892	2289	397
5	1.9	336	275	397	427	305	305	275	305	328	56	275	427	153
	4.1	458	671	763	732	671	641	671	641	656	91	458	763	305
	6.2	1495	1526	1373	1465	1495	1190	1465	1343	1419	112	1190	1526	336
	8.1	1800	2106	1526	1587	1587	1739	1770	1648	1720	184	1526	2106	580
	12.1	1984	2319	2136	1648	1953	2228	2228	2167	2083	215	1648	2319	671
	17.5	2197	2899	2167	2533	2868	3387	2624	2807	2685	400	2167	3387	1221
6	1.9	61	61	31	61	31	61	61	61	53	14	31	61	31
	4.1	214	214	275	244	153	214	153	153	202	46	153	275	122
	6.2	183	183	244	214	183	244	275	244	221	36	183	275	92
	8.1	244	275	305	244	275	244	275	275	267	22	244	305	61
	12.1	305	366	427	275	336	336	458	397	362	62	275	458	183
	17.5	671	488	610	458	580	580	580	610	572	69	458	671	214
7	1.0	153	214	183	183	183	153	214	183	183	23	153	214	61
	3.5	732	763	732	732	824	885	763	763	774	54	732	885	153
	7.7	1221	1312	1343	1129	1282	1190	1282	1282	1255	70	1129	1343	214
	11.5	1526	1556	1495	1312	1465	1495	1709	1709	1533	130	1312	1709	397
	16.9	2319	1678	1770	1953	1709	1739	1556	1556	1785	250	1556	2319	763
8	1.1	214	214	214	244	214	183	244	214	217	20	183	244	61
	3.8	580	580	610	610	519	519	580	549	568	36	519	610	92
	5.8	977	977	885	793	885	946	1007	1007	935	75	793	1007	214
	7.7	1007	1007	977	977	1251	1068	1282	1312	1110	146	977	1312	336
	9.9	1556	1556	1404	1404	1587	1617	1709	1709	1568	118	1404	1709	305
	11.8	2563	2106	2350	1984	2655	2411	2533	2685	2411	254	1984	2685	702

Table B2 - Measured Sample Mass Loss (pg. 2 of 6)

Site No.	Exp Time yrs	Mass Loss (g/m²) - 2 Sample Average Site								Average for Retrieval (g/m²)				
		Nom. 1.5 in. Pipe				Nom. 3.0 in. Pipe				Mean	Std. Dev.	Min.	Max.	Range
		a	b	e	y	B	K	M	Y					
9	1.0	183	366	214	275	183	183	183	153	217	70	153	366	214
	3.5	519	702	702	610	580	366	580	519	572	109	366	702	336
	5.5	671	702	732	702	793	824	824	793	755	60	671	824	153
	7.7	977	1038	977	915	1068	1038	977	946	992	52	915	1068	153
	11.5	1434	1465	1587	1526	1495	1526	1404	1434	1484	61	1404	1587	183
	16.9	1648	1831	1770	1648	1678	1648	1556	1587	1671	90	1556	1831	275
10	1.3	427	366	366	397	336	305	427	366	374	42	305	427	122
	4.0	427	641	824	366	519	366	793	488	553	181	366	824	458
	6.1	1038	1038	1068	1038	1038	1129	1038	1038	1053	33	1038	1129	92
	7.9	1343	1312	1373	1373	1434	1404	1404	1404	1381	39	1312	1434	122
	12.0	1312	1648	1495	1343	1465	1343	1495	1556	1457	117	1312	1648	336
11	1.4	92	153	122	153	122	122	122	122	126	20	92	153	61
	4.0	427	336	336	366	305	275	305	275	328	51	275	427	153
	6.0	336	427	427	397	397	427	397	458	408	36	336	458	122
	7.8	305	671	580	458	458	427	488	427	477	109	305	671	366
	10.0	427	519	610	610	519	549	488	458	523	66	427	610	183
	11.9	519	824	641	580	610	610	519	671	622	98	519	824	305
12	1.9	61	61	122	92	61	61	92	61	76	23	61	122	61
	4.1	427	580	488	610	458	458	397	488	488	73	397	610	214
	6.2	977	702	1038	915	1007	854	824	885	900	109	702	1038	336
	8.0	305	427	275	305	275	244	244	336	301	60	244	427	183
	12.1	1038	1343	1251	1190	1221	1099	1099	1007	1156	115	1007	1343	336
	17.5	2197	1709	1831	1709	1648	1465	1617	1800	1747	214	1465	2197	732
13	1.9	763	915	1129	1038	1190	1373	1007	1251	1083	194	763	1373	610
	4.2	824	2136	1190	1007	915	702	1343	1434	1194	457	702	2136	1434
	5.9	1984	2533	2045	2258	1678	1861	2380	2624	2170	333	1678	2624	946
14	1.1	92	122	122	153	122	153	122	122	126	20	92	153	61
	3.8	610	763	793	732	702	671	671	610	694	67	610	793	183
	5.8	732	1129	1099	946	1038	977	946	885	969	126	732	1129	397
	7.7	610	1129	763	671	580	519	1007	702	748	214	519	1129	610
	9.9	1160	1953	1373	1617	1526	1221	1404	1099	1419	280	1099	1953	854
	11.8	1251	1373	1587	1373	1465	1312	1404	1434	1400	101	1251	1587	336
15	2.0	732	641	702	641	671	702	824	824	717	73	641	824	183
	4.0	1099	915	915	977	946	1038	1160	946	999	90	915	1160	244
	5.9	1892	1861	1709	1648	1892	1739	1587	1953	1785	133	1587	1953	366
	8.0	1526	1434	1648	1678	1770	1739	1556	1526	1610	117	1434	1770	336
	12.0	2472	2594	2350	2380	2533	2502	2472	2777	2510	133	2350	2777	427
	17.6	3753	3784	3326	3174	3814	4883	3418	2319	3559	724	2319	4883	2563
16	2.0	580	763	824	641	671	702	793	641	702	85	580	824	244
	4.0	885	1190	1068	915	885	915	1007	1068	992	111	885	1190	305
	6.0	1587	1495	1343	1312	1709	1373	1526	1526	1484	134	1312	1709	397
	7.9	1556	1526	1404	1373	1404	1404	1373	1556	1449	82	1373	1556	183
	10.0	2045	2014	2075	1892	1770	1892	2106	1892	1961	116	1770	2106	336
	12.0	2533	2502	2441	2228	2350	2319	2289	2319	2373	108	2228	2533	305

Table B2 - Measured Sample Mass Loss (pg. 3 of 6)

Site No.	Exp Time yrs	Mass Loss (g/m²) - 2 Sample Average Site								Average for Retrieval (g/m²)				
		Nom. 1.5 in. Pipe				Nom. 3.0 in. Pipe				Mean	Std. Dev.	Min.	Max.	Range
		a	b	e	y	B	K	M	Y					
17	1.2	458	458	427	427	488	458	488	458	458	23	427	488	61
	3.8	1007	977	1038	977	1129	1038	1099	1099	1045	58	977	1129	153
	5.9	1587	1800	1709	1587	1861	1861	1831	1892	1766	124	1587	1892	305
	7.7	1953	2167	1953	2014	2258	2136	2258	2258	2125	134	1953	2258	305
	11.8	2868	2624	2533	2746	3235	3113	2930	3052	2888	244	2533	3235	702
	17.0	2411	2685	2899	2502	2899	2868	2899	2716	2735	193	2411	2899	488
18	1.2	122	214	122	183	122	92	153	153	145	39	92	214	122
	3.8	366	549	580	488	397	366	366	519	454	90	366	580	214
	5.8	702	1221	977	793	885	824	977	793	896	161	702	1221	519
	7.7	549	1007	885	610	671	580	702	641	706	159	549	1007	458
	9.8	854	1068	977	1007	915	1038	946	885	961	75	854	1068	214
	11.7	915	824	946	793	610	824	732	1190	854	171	610	1190	580
19	580	183	244	214	214	183	214	244	244	217	25	183	244	61
	1312	610	458	671	671	702	702	549	610	622	85	458	702	244
	1892	671	915	671	702	702	702	610	641	702	92	610	915	305
	2441	763	915	824	854	732	732	915	763	812	76	732	915	183
	3113	915	977	977	885	946	885	946	915	931	36	885	977	92
	3692	885	1068	1160	1038	977	1038	1007	1068	1030	79	885	1160	275
20	1221	275	244	336	305	275	305	275	275	286	28	244	336	92
	1861	671	732	549	702	671	671	641	610	656	57	549	732	183
	2411	732	824	732	641	702	702	854	671	732	73	641	854	214
	3662	824	1038	824	763	977	885	1068	854	904	111	763	1068	305
	5249	1190	1465	1404	1251	1556	1404	1465	1282	1377	125	1190	1556	366
	305	2014	1984	1861	1831	1709	1861	1831	1709	1850	111	1709	2014	305
21	1.5	610	641	641	641	671	610	610	641	633	22	610	671	61
	4.0	763	793	885	732	854	885	885	763	820	64	732	885	153
	6.0	1282	1343	1526	1434	1465	1556	1404	1404	1427	90	1282	1556	275
22	1.7	427	488	610	580	427	458	427	549	496	74	427	610	183
	3.7	977	1190	1038	1129	1099	1129	1068	1068	1087	65	977	1190	214
	5.6	1617	1648	1404	1648	1404	1404	1373	1099	1449	186	1099	1648	549
	7.6	1739	1892	2258	1770	1953	1892	2380	2289	2022	250	1739	2380	641
	9.6	2045	2136	2289	1953	2197	2075	2289	1984	2121	129	1953	2289	336
	11.6	2136	2319	2289	2167	2411	2106	2380	2136	2243	121	2106	2411	305
23	1.9	1648	2319	2350	2380	1770	1984	2045	2380	2109	292	1648	2380	732
	4.3	3418	4059	4181	3204	3662	3631	3662	2960	3597	406	2960	4181	1221
	6.2	4181	4791	3723	4791	3937	4425	4211	5310	4421	519	3723	5310	1587
	8.0	5554	6256	4730	5676	5676	6439	5859	7263	5931	742	4730	7263	2533
	10.2	5951	6591	5890	5767	6012	6012	6622	6164	6126	317	5767	6622	854
	12.1	6134	6439	6042	6225	6317	5981	5981	7415	6317	473	5981	7415	1434
24	1.3	61	92	92	92	61	92	92	61	80	16	61	92	31
	4.0	122	183	153	122	122	122	122	92	130	27	92	183	92
	6.1	275	366	397	244	336	336	336	305	324	49	244	397	153
	7.9	275	305	305	244	305	275	305	244	282	27	244	305	61
	12.0	397	427	458	427	366	427	427	366	412	33	366	458	92
	17.2	397	580	549	427	427	397	488	366	454	77	366	580	214

Table B2 - Measured Sample Mass Loss (pg. 4 of 6)

Site No.	Exp Time yrs	Mass Loss (g/m²) - 2 Sample Average Site								Average for Retrieval (g/m²)				
		Nom. 1.5 in. Pipe				Nom. 3.0 in. Pipe				Mean	Std. Dev.	Min.	Max.	Range
		a	b	e	y	B	K	M	Y					
25	1.0	122	183	122	122	92	122	122	122	126	25	92	183	92
	3.7	183	427	336	336	214	275	427	427	328	98	183	427	244
	5.7	519	519	580	580	549	519	519	458	530	40	458	580	122
	7.6	549	702	610	610	580	549	549	519	584	58	519	702	183
	11.7	1038	1221	1038	885	854	1099	1007	854	999	129	854	1221	366
	17.0	915	1282	1160	946	854	977	824	824	973	166	824	1282	458
26	1.0	244	275	244	244	244	244	275	275	256	16	244	275	31
	3.5	488	671	549	549	580	580	671	580	584	62	488	671	183
	5.5	458	641	519	458	366	366	427	427	458	89	366	641	275
	7.7	397	641	610	488	458	458	488	519	507	81	397	641	244
	11.5	1068	1282	1099	1068	1221	1129	1160	1038	1133	84	1038	1282	244
	16.9	1221	1556	1312	1251	1434	1251	1465	1312	1350	121	1221	1556	336
27	2.0	122	183	153	153	183	183	183	153	164	23	122	183	61
	4.0	977	946	1190	1099	1068	1068	1007	824	1022	111	824	1190	366
	6.0	915	1160	1251	915	946	946	915	977	1003	129	915	1251	336
	8.0	1404	1312	1282	1648	1312	1495	1190	1343	1373	142	1190	1648	458
	12.0	2167	2594	2685	2319	2533	2136	2502	2472	2426	199	2136	2685	549
	17.6	2838	3479	3082	2838	2960	2594	2716	2563	2884	297	2563	3479	915
28	1.6	1038	824	1099	915	641	977	885	1068	931	150	641	1099	458
	5.6	3143	2838	2899	3235	2716	3265	3204	3265	3071	219	2716	3265	549
	7.7	3540	3906	4425	4608	4669	5005	4028	5584	4471	651	3540	5584	2045
	9.6	4791	4699	5005	5127	5127	4944	4730	5401	4978	239	4699	5401	702
29	2.0	1221	1007	1160	1343	1068	1282	1251	1282	1202	115	1007	1343	336
	4.1	2197	2350	2228	2075	2197	2228	2106	2411	2224	112	2075	2411	336
	6.0	3052	2716	2868	3021	2807	2930	2716	2838	2868	126	2716	3052	336
	8.0	4364	3967	4181	4272	3326	5005	3357	4791	4158	603	3326	5005	1678
	10.0	4944	4089	4211	4272	4547	4669	4516	4760	4501	293	4089	4944	854
	12.0	6561	4944	5859	5920	4760	5462	4516	6134	5520	723	4516	6561	2045
30	1.1	275	275	336	275	275	336	336	305	301	30	275	336	61
	3.6	336	366	366	366	427	397	427	366	381	33	336	427	92
	5.7	549	702	793	610	763	824	763	732	717	94	549	824	275
	8.2	1251	1434	1251	1282	1434	1221	1343	1160	1297	99	1160	1434	275
	11.6	1587	1709	1465	1617	1922	1770	1709	1587	1671	139	1465	1922	458
	17.0	1861	1800	1739	1648	1831	1831	2106	1953	1846	137	1648	2106	458
31	2.0	214	397	366	458	244	366	397	336	347	81	214	458	244
	4.1	610	519	641	671	641	763	641	702	648	71	519	763	244
	6.0	488	549	580	580	458	580	702	702	580	88	458	702	244
	8.0	549	671	732	854	641	671	702	793	702	94	549	854	305
	12.0	885	915	854	824	854	854	854	1038	885	67	824	1038	214
	17.7	1495	1282	1129	1343	1922	1373	1251	1404	1400	238	1129	1922	793
32	1.0	61	153	122	122	92	122	122	92	111	28	61	153	92
	3.7	366	549	488	427	427	397	397	427	435	58	366	549	183
	5.8	549	641	702	549	702	671	793	763	671	89	549	793	244
	7.6	610	854	824	702	763	793	702	824	759	82	610	854	244
	9.6	641	977	946	763	915	732	854	793	828	116	641	977	336
	11.7	946	1068	1007	1129	1099	1099	1282	977	1076	105	946	1282	336

Table B2 - Measured Sample Mass Loss (pg. 5 of 6)

Site No.	Exp Time yrs	Mass Loss (g/m²) - 2 Sample Average Site								Average for Retrieval (g/m²)				
		Nom. 1.5 in. Pipe				Nom. 3.0 in. Pipe				Mean	Std. Dev.	Min.	Max.	Range
		a	b	e	y	B	K	M	Y					
33	1.0	122	122	122	153	122	122	122	122	126	11	122	153	31
	3.7	1068	1007	1038	1068	1129	1068	1099	1160	1079	49	1007	1160	153
	5.8	1434	1404	1404	1709	1312	1709	1465	915	1419	250	915	1709	793
	7.6	2563	2807	2807	2685	2594	2685	2716	2380	2655	141	2380	2807	427
	9.7	3387	3784	3479	2991	3082	3021	3479	3204	3303	277	2991	3784	793
	11.7	4303	4272	4364	4333	4364	3540	3906	4272	4169	294	3540	4364	824
34	1.4	397	397	366	336	427	427	427	366	393	34	336	427	92
	4.0	488	549	549	580	549	580	549	549	549	28	488	580	92
	6.1	946	854	854	824	854	946	854	885	877	45	824	946	122
	8.0	915	1007	1038	1068	1068	977	1007	977	1007	52	915	1068	153
	9.9	1282	1465	1404	1251	1373	1190	1312	1343	1327	88	1190	1465	275
	12.0	1160	1465	1221	977	1800	1312	1617	1465	1377	264	977	1800	824
35	1.9	244	275	214	214	244	275	336	275	259	40	214	336	122
	4.1	519	610	549	427	641	610	580	549	561	67	427	641	214
	6.2	366	366	458	305	2746	214	305	427	648	851	214	2746	2533
	8.0	732	641	519	519	641	641	641	702	629	76	519	732	214
	12.1	488	824	671	336	549	519	763	580	591	157	336	824	488
	17.5	458	1251	458	275	366	275	977	1007	633	383	275	1251	977
36	2.0	366	366	427	336	244	244	305	275	320	65	244	427	183
	4.1	427	610	671	641	519	519	488	458	542	89	427	671	244
	6.0	397	519	519	549	427	549	427	488	484	60	397	549	153
	8.0	610	915	763	732	549	610	641	641	683	116	549	915	366
	12.0	763	1007	977	885	702	854	732	763	835	114	702	1007	305
	17.7	1007	1251	1251	1129	763	915	793	946	1007	190	763	1251	488
37	2.0	702	702	732	732	732	732	763	732	729	20	702	763	61
	4.1	1282	1343	1343	1282	1373	1343	1373	1526	1358	77	1282	1526	244
	6.0	1282	1404	1099	1404	1160	1282	1251	1160	1255	112	1099	1404	305
	8.0	1770	1861	1587	1587	1922	1861	1770	1984	1793	146	1587	1984	397
	10.1	2777	2624	2441	2350	2502	2655	2594	3052	2624	218	2350	3052	702
	12.0	2533	2746	2350	2075	2502	2563	2746	3174	2586	322	2075	3174	1099
38	1.4	61	61	61	61	61	92	92		72	16	61	92	31
	4.0	153	214	214	183	153	122	183	153	172	32	122	214	92
	6.1	275	244	366	244	336	397	366	366	324	61	244	397	153
	8.0	610	610	641	549	610	610	671	641	618	36	549	671	122
	12.0	641	732	854	793	732	763	671	824	751	73	641	854	214
	17.2	824	1038	702	793	702	854	854	763	816	108	702	1038	336
39	1.4	305	366	366	397	366	397	366	641	401	101	305	641	336
	4.0	946	854	885	824	977	915	977	885	908	56	824	977	153
	6.1	702	977	793	824	977	946	977	977	896	108	702	977	275
	8.0	1251	1373	1404	1099	1495	1465	1495	1556	1392	151	1099	1556	458
	9.9	1404	1587	1495	1373	1617	1526	1434	1465	1488	86	1373	1617	244
	12.0	1465	1678	1861	1587	1922	1709	1953	1526	1713	184	1465	1953	488
40	2.0	641	641	641	610	702	763	671	854	690	81	610	854	244
	4.1	1221	1312	1190	1129	1282	1312	1404	1282	1266	85	1129	1404	275
	6.0	1648	1678	1526	1617	1861	2014	1861	1709	1739	160	1526	2014	488
	8.0	1953	1984	1984	1678	2136	2167	2502	2350	2094	255	1678	2502	824
	10.0	2106	2197	2228	1831	2472	2197	2838	2289	2270	292	1831	2838	1007
	12.0	2106	2411	2197	2045	2624	2624	2136	2533	2334	241	2045	2624	580

Table B2 - Measured Sample Mass Loss (pg. 6 of 6)

Site No.	Exp Time yrs	Mass Loss (g/m²) - 2 Sample Average Site								Average for Retrieval (g/m²)				
		Nom. 1.5 in. Pipe				Nom. 3.0 in. Pipe				Mean	Std. Dev.	Min.	Max.	Range
		a	b	e	y	B	K	M	Y					
41	1.5	244	153	153	305	183	183	122	122	183	63	122	305	183
	4.0	824	1099	915	915	824	885	732	763	870	114	732	1099	366
	6.0	1404	1434	1373	1282	977	1099	1251	1190	1251	158	977	1434	458
	7.9	1465	1648	1526	1434	1282	1282	1404	1343	1423	125	1282	1648	366
	12.0	1831	1892	1922	1617	1984	1922	1831	1831	1854	110	1617	1984	366
	17.4	1800	2258	2106	2136	1678	1678	2136	2136	1991	233	1678	2258	580
42	2.0	915	1099	1007	946	885	732	885	580	881	161	580	1099	519
	4.1	1373	2228	1739	1800	1465	1038	1556	1038	1530	400	1038	2228	1190
	6.0	2136	2441	2350	2228	1739	2106	1892	2014	2113	232	1739	2441	702
	8.0	1800	2350	2502	2838	1953	2075	2045	1709	2159	380	1709	2838	1129
	10.1	3509	3235	3692	3814	2685	3387	2838	2807	3246	428	2685	3814	1129
	12.0	3235	5218	3814	5310	2716	3418	3357	4089	3895	937	2716	5310	2594
43	1.3	427	519	580	854	610	671	702	732	637	133	427	854	427
	4.1	1099	2197	1404	1465	1434	1587	2106	1953	1655	387	1099	2197	1099
	6.2	1770	2441	2960	2258	2197	1892	2350	2533	2300	373	1770	2960	1190
	8.0	2777	4669	3143	3479	3082	3296	2746	4394	3448	715	2746	4669	1922
	9.9	3540	3479	3204	5188	3174	3418	2991	3662	3582	685	2991	5188	2197
	12.0	4730	5066	5951	5371	4303	3875	3998	4974	4783	707	3875	5951	2075
44	1.1	92	183	122	153	122	92	122	122	126	30	92	183	92
	3.6	427	549	610	549	427	427	366	397	469	88	366	610	244
	5.7	702	671	702	732	641	671	610	610	668	44	610	732	122
	7.6	519	702	671	610	580	641	610	641	622	56	519	702	183
	11.6	885	1251	1434	1068	1038	854	1038	977	1068	192	854	1434	580
45	1.2	366	427	427	397	336	305	519	427	401	66	305	519	214
	3.8	1099	915	885	1007	885	977	854	1007	954	83	854	1099	244
	5.8	885	977	702	946	1007	946	977	915	919	96	702	1007	305
	7.7	1068	1709	1282	1160	1251	1190	1160	1282	1263	194	1068	1709	641
	9.8	4181	3631	3631	3692	3814	3998	3753	4120	3853	219	3631	4181	549
	11.7	2960	2746	2960	2838	3235	3448	2838	3418	3055	274	2746	3448	702
46	1.5	244	397	366	275	305	336	366	366	332	53	244	397	153
	4.0	824	977	885	793	732	793	824	977	851	88	732	977	244
	5.1	885	854	946	915	854	915	793	977	893	58	793	977	183
	8.0	1709	1892	1587	1739	1739	1800	1770	2045	1785	135	1587	2045	458
	10.2	1221	1434	1465	1251	1343	1099	1312	1190	1289	124	1099	1465	366
	12.0	1221	1556	1373	1343	1434	1312	1465	1465	1396	105	1221	1556	336
47	1.5	153	153	1526	122	122	183	153	122	317	489	122	1526	1404
	4.1	397	580	519	397	397	397	488	610	473	89	397	610	214
	6.1	366	458	488	366	580	488	671	458	484	102	366	671	305
	8.0	671	641	519	549	641	641	610	488	595	67	488	671	183
	12.1	702	1190	915	854	977	915	854	946	919	138	702	1190	488
	17.4	1770	1953	2411	2563	1861	1984	1648	1556	1968	353	1556	2563	1007

Table B3 - Mass Loss Rates for Sites (pg. 1 of 6)

Site No.	Exp Time yrs	Mass Loss Rate (g/m²/d) - 2 Sample Average								Site Average (g/m²/d)				
		Nom. 1.5 in. Pipe				Nom. 3.0 in. Pipe				Mean	Std. Dev.	Min.	Max.	Range
		a	b	e	y	B	K	M	Y					
1	1.0	0.92	1.00	0.92	0.92	1.00	1.17	1.09	1.00	0.81	0.14	0.50	1.17	0.67
	3.6	0.77	0.93	0.91	0.91	0.97	0.97	1.00	0.88					
	5.5	0.67	0.84	0.84	0.71	0.82	0.82	0.77	0.70					
	7.7	0.54	0.74	0.74	0.72	0.72	0.61	0.80	0.72					
	9.6	0.70	0.82	0.84	0.77	0.82	0.88	0.77	0.79					
	11.6	0.64	0.63	0.50	0.67	0.68	0.68	0.84	0.63					
2	2.1	0.76	0.80	0.72	0.95	0.76	0.88	0.88	0.72	0.52	0.17	0.24	0.95	0.71
	4.0	0.61	0.63	0.63	0.63	0.73	0.56	0.65	0.52					
	5.9	0.54	0.59	0.55	0.48	0.51	0.57	0.59	0.59					
	7.9	0.40	0.39	0.39	0.38	0.39	0.33	0.41	0.24					
	12.0	0.42	0.49	0.40	0.41	0.44	0.38	0.46	0.41					
	17.6	0.37	0.40	0.37	0.38	0.38	0.37	0.34	0.35					
3	2.0	0.63	0.75	0.71	0.84	0.67	0.84	0.88	0.71	0.50	0.18	0.26	0.88	0.62
	4.1	0.59	0.65	0.67	0.71	0.73	0.67	0.73	0.79					
	6.0	0.45	0.58	0.52	0.49	0.53	0.50	0.50	0.58					
	8.0	0.39	0.46	0.44	0.41	0.39	0.43	0.47	0.40					
	10.1	0.35	0.38	0.36	0.34	0.26	0.30	0.31	0.33					
	12.1	0.26	0.34	0.32	0.35	0.32	0.29	0.30	0.31					
4	1.4	0.90	0.72	0.90	0.84	0.84	0.84	0.84	0.90	0.65	0.13	0.43	0.90	0.46
	4.0	0.77	0.73	0.73	0.73	0.77	0.77	0.73	0.77					
	6.1	0.62	0.60	0.60	0.63	0.64	0.63	0.60	0.67					
	8.0	0.54	0.56	0.56	0.55	0.57	0.57	0.56	0.61					
	12.0	0.49	0.46	0.49	0.43	0.49	0.49	0.48	0.52					
5	1.9	0.48	0.40	0.57	0.62	0.44	0.44	0.40	0.44	0.50	0.10	0.31	0.71	0.41
	4.1	0.31	0.45	0.51	0.49	0.45	0.43	0.45	0.43					
	6.2	0.66	0.67	0.61	0.65	0.66	0.53	0.65	0.59					
	8.1	0.61	0.71	0.52	0.54	0.54	0.59	0.60	0.56					
	12.1	0.45	0.52	0.48	0.37	0.44	0.50	0.50	0.49					
	17.5	0.34	0.45	0.34	0.40	0.45	0.53	0.41	0.44					
6	1.9	0.09	0.09	0.04	0.09	0.04	0.09	0.09	0.09	0.10	0.03	0.04	0.18	0.14
	4.1	0.14	0.14	0.18	0.16	0.10	0.14	0.10	0.10					
	6.2	0.08	0.08	0.11	0.09	0.08	0.11	0.12	0.11					
	8.1	0.08	0.09	0.10	0.08	0.09	0.08	0.09	0.09					
	12.1	0.07	0.08	0.10	0.06	0.08	0.08	0.10	0.09					
	17.5	0.11	0.08	0.10	0.07	0.09	0.09	0.09	0.10					
7	1.0	0.42	0.58	0.50	0.50	0.50	0.42	0.58	0.50	0.44	0.12	0.25	0.69	0.44
	3.5	0.57	0.60	0.57	0.57	0.64	0.69	0.60	0.60					
	7.7	0.43	0.47	0.48	0.40	0.46	0.42	0.46	0.46					
	11.5	0.36	0.37	0.36	0.31	0.35	0.36	0.41	0.41					
	16.9	0.38	0.27	0.29	0.32	0.28	0.28	0.25	0.25					
8	1.1	0.53	0.53	0.53	0.61	0.53	0.46	0.61	0.53	0.46	0.08	0.35	0.62	0.28
	3.8	0.42	0.42	0.44	0.44	0.37	0.37	0.42	0.40					
	5.8	0.46	0.46	0.42	0.37	0.42	0.45	0.48	0.48					
	7.7	0.36	0.36	0.35	0.35	0.44	0.38	0.46	0.47					
	9.9	0.43	0.43	0.39	0.39	0.44	0.45	0.47	0.47					
	11.8	0.59	0.49	0.55	0.46	0.62	0.56	0.59	0.62					

Table B3 - Mass Loss Rates for Sites (pg. 2 of 6)

Site No.	Exp Time yrs	Mass Loss Rate (g/m²/d) - 2 Sample Average								Site Average (g/m²/d)				
		Nom. 1.5 in. Pipe				Nom. 3.0 in. Pipe				Mean	Std. Dev.	Min.	Max.	Range
		a	b	e	y	B	K	M	Y					
9	1.0	0.50	1.00	0.58	0.75	0.50	0.50	0.50	0.42	0.40	0.13	0.25	1.00	0.75
	3.5	0.41	0.55	0.55	0.48	0.45	0.29	0.45	0.41					
	5.5	0.33	0.35	0.36	0.35	0.39	0.41	0.41	0.39					
	7.7	0.35	0.37	0.35	0.33	0.38	0.37	0.35	0.34					
	11.5	0.34	0.35	0.38	0.36	0.36	0.36	0.33	0.34					
	16.9	0.27	0.30	0.29	0.27	0.27	0.27	0.25	0.26					
10	1.3	0.90	0.77	0.77	0.84	0.71	0.64	0.90	0.77	0.49	0.17	0.25	0.90	0.65
	4.0	0.29	0.44	0.56	0.25	0.36	0.25	0.54	0.33					
	6.1	0.47	0.47	0.48	0.47	0.47	0.51	0.47	0.47					
	7.9	0.47	0.45	0.48	0.48	0.50	0.49	0.49	0.49					
	12.0	0.30	0.38	0.34	0.31	0.33	0.31	0.34	0.36					
11	1.4	0.18	0.30	0.24	0.30	0.24	0.24	0.24	0.24	0.19	0.05	0.11	0.30	0.19
	4.0	0.29	0.23	0.23	0.25	0.21	0.19	0.21	0.19					
	6.0	0.15	0.19	0.19	0.18	0.18	0.19	0.18	0.21					
	7.8	0.11	0.24	0.20	0.16	0.16	0.15	0.17	0.15					
	10.0	0.12	0.14	0.17	0.17	0.14	0.15	0.13	0.13					
	11.9	0.12	0.19	0.15	0.13	0.14	0.14	0.12	0.15					
12	1.9	0.09	0.09	0.18	0.13	0.09	0.09	0.13	0.09	0.25	0.11	0.08	0.46	0.37
	4.1	0.29	0.39	0.33	0.41	0.31	0.31	0.26	0.33					
	6.2	0.43	0.31	0.46	0.40	0.44	0.38	0.36	0.39					
	8.0	0.10	0.15	0.09	0.10	0.09	0.08	0.08	0.11					
	12.1	0.23	0.30	0.28	0.27	0.28	0.25	0.25	0.23					
	17.5	0.34	0.27	0.29	0.27	0.26	0.23	0.25	0.28					
13	1.9	1.10	1.32	1.63	1.50	1.71	1.98	1.45	1.80	1.12	0.41	0.46	1.98	1.52
	4.2	0.54	1.39	0.78	0.66	0.60	0.46	0.88	0.93					
	5.9	0.92	1.18	0.95	1.05	0.78	0.86	1.10	1.22					
14	1.1	0.23	0.30	0.30	0.38	0.30	0.38	0.30	0.30	0.38	0.10	0.18	0.57	0.39
	3.8	0.44	0.55	0.57	0.53	0.51	0.48	0.48	0.44					
	5.8	0.35	0.53	0.52	0.45	0.49	0.46	0.45	0.42					
	7.7	0.22	0.40	0.27	0.24	0.21	0.18	0.36	0.25					
	9.9	0.32	0.54	0.38	0.45	0.42	0.34	0.39	0.30					
	11.8	0.29	0.32	0.37	0.32	0.34	0.30	0.33	0.33					
15	2.0	1.00	0.88	0.96	0.88	0.92	0.96	1.13	1.13	0.70	0.18	0.36	1.13	0.77
	4.0	0.75	0.63	0.63	0.67	0.65	0.71	0.79	0.65					
	5.9	0.88	0.86	0.79	0.76	0.88	0.81	0.74	0.91					
	8.0	0.52	0.49	0.56	0.57	0.61	0.60	0.53	0.52					
	12.0	0.56	0.59	0.54	0.54	0.58	0.57	0.56	0.63					
	17.6	0.58	0.59	0.52	0.49	0.59	0.76	0.53	0.36					
16	2.0	0.79	1.04	1.13	0.88	0.92	0.96	1.09	0.88	0.65	0.17	0.48	1.13	0.65
	4.0	0.61	0.81	0.73	0.63	0.61	0.63	0.69	0.73					
	6.0	0.72	0.68	0.61	0.60	0.78	0.63	0.70	0.70					
	7.9	0.54	0.53	0.49	0.48	0.49	0.49	0.48	0.54					
	10.0	0.56	0.55	0.57	0.52	0.48	0.52	0.58	0.52					
	12.0	0.58	0.57	0.56	0.51	0.54	0.53	0.52	0.53					

Table B3 - Mass Loss Rates for Sites (pg. 3 of 6)

Site No.	Exp Time yrs	Mass Loss Rate (g/m²/d) - 2 Sample Average								Site Average (g/m²/d)				
		Nom. 1.5 in. Pipe				Nom. 3.0 in. Pipe				Mean	Std. Dev.	Min.	Max.	Range
		a	b	e	y	B	K	M	Y					
17	1.2	1.04	1.04	0.97	0.97	1.11	1.04	1.11	1.04	0.75	0.19	0.39	1.11	0.73
	3.8	0.73	0.70	0.75	0.70	0.81	0.75	0.79	0.79					
	5.9	0.74	0.84	0.79	0.74	0.86	0.86	0.85	0.88					
	7.7	0.69	0.77	0.69	0.72	0.80	0.76	0.80	0.80					
	11.8	0.67	0.61	0.59	0.64	0.75	0.72	0.68	0.71					
	17.0	0.39	0.43	0.47	0.40	0.47	0.46	0.47	0.44					
18	1.2	0.28	0.49	0.28	0.42	0.28	0.21	0.35	0.35	0.30	0.09	0.14	0.58	0.43
	3.8	0.26	0.40	0.42	0.35	0.29	0.26	0.26	0.37					
	5.8	0.33	0.58	0.46	0.37	0.42	0.39	0.46	0.37					
	7.7	0.20	0.36	0.31	0.22	0.24	0.21	0.25	0.23					
	9.8	0.24	0.30	0.27	0.28	0.26	0.29	0.26	0.25					
	11.7	0.21	0.19	0.22	0.19	0.14	0.19	0.17	0.28					
19	1.1	0.46	0.61	0.53	0.53	0.46	0.53	0.61	0.61	0.36	0.12	0.21	0.61	0.40
	3.7	0.45	0.34	0.50	0.50	0.52	0.52	0.41	0.45					
	5.7	0.32	0.44	0.32	0.34	0.34	0.34	0.29	0.31					
	7.6	0.27	0.33	0.30	0.31	0.26	0.26	0.33	0.27					
	9.7	0.26	0.28	0.28	0.25	0.27	0.25	0.27	0.26					
	11.6	0.21	0.25	0.27	0.24	0.23	0.24	0.24	0.25					
20	1.0	0.75	0.67	0.92	0.84	0.75	0.84	0.75	0.75	0.47	0.16	0.27	0.92	0.65
	3.6	0.51	0.56	0.42	0.53	0.51	0.51	0.49	0.46					
	5.5	0.36	0.41	0.36	0.32	0.35	0.35	0.43	0.33					
	7.7	0.29	0.37	0.29	0.27	0.35	0.31	0.38	0.30					
	9.6	0.34	0.42	0.40	0.36	0.44	0.40	0.42	0.37					
	11.6	0.48	0.47	0.44	0.43	0.40	0.44	0.43	0.40					
21	1.5	1.11	1.17	1.17	1.17	1.23	1.11	1.11	1.17	0.79	0.27	0.50	1.23	0.72
	4.0	0.52	0.54	0.61	0.50	0.58	0.61	0.61	0.52					
	6.0	0.58	0.61	0.70	0.65	0.67	0.71	0.64	0.64					
22	1.7	0.69	0.79	0.98	0.93	0.69	0.74	0.69	0.88	0.70	0.12	0.50	0.98	0.49
	3.7	0.72	0.88	0.77	0.84	0.81	0.84	0.79	0.79					
	5.6	0.79	0.81	0.69	0.81	0.69	0.69	0.67	0.54					
	7.6	0.63	0.68	0.81	0.64	0.70	0.68	0.86	0.82					
	9.6	0.58	0.61	0.65	0.56	0.63	0.59	0.65	0.57					
	11.6	0.50	0.55	0.54	0.51	0.57	0.50	0.56	0.50					
23	1.9	2.37	3.34	3.39	3.43	2.55	2.86	2.95	3.43	2.06	0.57	1.35	3.43	2.08
	4.3	2.18	2.58	2.66	2.04	2.33	2.31	2.33	1.88					
	6.2	1.85	2.12	1.64	2.12	1.74	1.95	1.86	2.34					
	8.0	1.90	2.14	1.62	1.94	1.94	2.20	2.01	2.49					
	10.2	1.60	1.77	1.58	1.55	1.61	1.61	1.78	1.65					
	12.1	1.39	1.46	1.37	1.41	1.43	1.35	1.35	1.68					
24	1.3	0.13	0.19	0.19	0.19	0.13	0.19	0.19	0.13	0.11	0.04	0.06	0.19	0.13
	4.0	0.08	0.13	0.10	0.08	0.08	0.08	0.08	0.06					
	6.1	0.12	0.16	0.18	0.11	0.15	0.15	0.15	0.14					
	7.9	0.10	0.11	0.11	0.08	0.11	0.10	0.11	0.08					
	12.0	0.09	0.10	0.10	0.10	0.08	0.10	0.10	0.08					
	17.2	0.06	0.09	0.09	0.07	0.07	0.06	0.08	0.06					

Table B3 - Mass Loss Rates for Sites (pg. 4 of 6)

Site No.	Exp Time yrs	Mass Loss Rate (g/m^2/d) - 2 Sample Average								Site Average (g/m^2/d)				
		Nom. 1.5 in. Pipe				Nom. 3.0 in. Pipe				Mean	Std. Dev.	Min.	Max.	Range
		a	b	e	y	B	K	M	Y					
25	1.0	0.33	0.50	0.33	0.33	0.25	0.33	0.33	0.33	0.24	0.07	0.13	0.50	0.37
	3.7	0.14	0.32	0.25	0.25	0.16	0.20	0.32	0.32					
	5.7	0.25	0.25	0.28	0.28	0.26	0.25	0.25	0.22					
	7.6	0.20	0.25	0.22	0.22	0.21	0.20	0.20	0.19					
	11.7	0.24	0.29	0.24	0.21	0.20	0.26	0.24	0.20					
	17.0	0.15	0.21	0.19	0.15	0.14	0.16	0.13	0.13					
26	1.0	0.67	0.75	0.67	0.67	0.67	0.67	0.75	0.75	0.34	0.19	0.14	0.75	0.61
	3.5	0.38	0.53	0.43	0.43	0.45	0.45	0.53	0.45					
	5.5	0.23	0.32	0.26	0.23	0.18	0.18	0.21	0.21					
	7.7	0.14	0.23	0.22	0.17	0.16	0.16	0.17	0.18					
	11.5	0.25	0.31	0.26	0.25	0.29	0.27	0.28	0.25					
	16.9	0.20	0.25	0.21	0.20	0.23	0.20	0.24	0.21					
27	2.0	0.17	0.25	0.21	0.21	0.25	0.25	0.25	0.21	0.48	0.15	0.17	0.81	0.65
	4.0	0.67	0.65	0.81	0.75	0.73	0.73	0.69	0.56					
	6.0	0.42	0.53	0.57	0.42	0.43	0.43	0.42	0.45					
	8.0	0.48	0.45	0.44	0.56	0.45	0.51	0.41	0.46					
	12.0	0.49	0.59	0.61	0.53	0.58	0.49	0.57	0.56					
	17.6	0.44	0.54	0.48	0.44	0.46	0.40	0.42	0.40					
28	1.6	1.78	1.41	1.88	1.57	1.10	1.67	1.51	1.83	1.53	0.19	1.10	1.99	0.89
	5.6	1.54	1.39	1.42	1.58	1.33	1.60	1.57	1.60					
	7.7	1.26	1.39	1.57	1.64	1.66	1.78	1.43	1.99					
	9.6	1.37	1.34	1.43	1.46	1.46	1.41	1.35	1.54					
29	2.0	1.67	1.38	1.59	1.84	1.46	1.75	1.71	1.75	1.39	0.19	1.03	1.84	0.81
	4.1	1.47	1.57	1.49	1.39	1.47	1.49	1.41	1.61					
	6.0	1.39	1.24	1.31	1.38	1.28	1.34	1.24	1.30					
	8.0	1.49	1.36	1.43	1.46	1.14	1.71	1.15	1.64					
	10.0	1.35	1.12	1.15	1.17	1.24	1.28	1.24	1.30					
	12.0	1.50	1.13	1.34	1.35	1.09	1.25	1.03	1.40					
30	1.1	0.68	0.68	0.84	0.68	0.68	0.84	0.84	0.76	0.42	0.16	0.26	0.84	0.58
	3.6	0.26	0.28	0.28	0.28	0.32	0.30	0.32	0.28					
	5.7	0.26	0.34	0.38	0.29	0.37	0.40	0.37	0.35					
	8.2	0.42	0.48	0.42	0.43	0.48	0.41	0.45	0.39					
	11.6	0.37	0.40	0.35	0.38	0.45	0.42	0.40	0.37					
	17.0	0.30	0.29	0.28	0.27	0.29	0.29	0.34	0.31					
31	2.0	0.29	0.54	0.50	0.63	0.33	0.50	0.54	0.46	0.31	0.12	0.17	0.63	0.45
	4.1	0.41	0.35	0.43	0.45	0.43	0.51	0.43	0.47					
	6.0	0.22	0.25	0.26	0.26	0.21	0.26	0.32	0.32					
	8.0	0.19	0.23	0.25	0.29	0.22	0.23	0.24	0.27					
	12.0	0.20	0.21	0.19	0.19	0.19	0.19	0.19	0.24					
	17.7	0.23	0.20	0.17	0.21	0.30	0.21	0.19	0.22					
32	1.0	0.17	0.42	0.33	0.33	0.25	0.33	0.33	0.25	0.28	0.05	0.17	0.42	0.25
	3.7	0.27	0.41	0.36	0.32	0.32	0.29	0.29	0.32					
	5.8	0.26	0.30	0.33	0.26	0.33	0.32	0.37	0.36					
	7.6	0.22	0.31	0.30	0.25	0.27	0.29	0.25	0.30					
	9.6	0.18	0.28	0.27	0.22	0.26	0.21	0.24	0.23					
	11.7	0.22	0.25	0.24	0.26	0.26	0.26	0.30	0.23					

Table B3 - Mass Loss Rates for Sites (pg. 5 of 6)

Site No.	Exp Time yrs	Mass Loss Rate (g/m^2/d) - 2 Sample Average								Site Average (g/m^2/d)				
		Nom. 1.5 in. Pipe				Nom. 3.0 in. Pipe				Mean	Std. Dev.	Min.	Max.	Range
		a	b	e	y	B	K	M	Y					
33	1.0	0.33	0.33	0.33	0.42	0.33	0.33	0.33	0.33	0.78	0.23	0.33	1.07	0.73
	3.7	0.79	0.75	0.77	0.79	0.84	0.79	0.81	0.86					
	5.8	0.68	0.66	0.66	0.81	0.62	0.81	0.69	0.43					
	7.6	0.92	1.01	1.01	0.97	0.93	0.97	0.98	0.86					
	9.7	0.96	1.07	0.98	0.84	0.87	0.85	0.98	0.90					
	11.7	1.01	1.00	1.02	1.01	1.02	0.83	0.91	1.00					
34	1.4	0.78	0.78	0.72	0.66	0.84	0.84	0.84	0.72	0.43	0.16	0.22	0.84	0.61
	4.0	0.33	0.38	0.38	0.40	0.38	0.40	0.38	0.38					
	6.1	0.42	0.38	0.38	0.37	0.38	0.42	0.38	0.40					
	8.0	0.31	0.34	0.36	0.37	0.37	0.33	0.34	0.33					
	9.9	0.35	0.41	0.39	0.35	0.38	0.33	0.36	0.37					
	12.0	0.26	0.33	0.28	0.22	0.41	0.30	0.37	0.33					
35	1.9	0.35	0.40	0.31	0.31	0.35	0.40	0.48	0.40	0.25	0.19	0.04	1.21	1.17
	4.1	0.35	0.41	0.37	0.29	0.43	0.41	0.39	0.37					
	6.2	0.16	0.16	0.20	0.13	1.21	0.09	0.13	0.19					
	8.0	0.25	0.22	0.18	0.18	0.22	0.22	0.22	0.24					
	12.1	0.11	0.19	0.15	0.08	0.12	0.12	0.17	0.13					
	17.5	0.07	0.20	0.07	0.04	0.06	0.04	0.15	0.16					
36	2.0	0.50	0.50	0.58	0.46	0.33	0.33	0.42	0.38	0.27	0.11	0.12	0.58	0.47
	4.1	0.29	0.41	0.45	0.43	0.35	0.35	0.33	0.31					
	6.0	0.18	0.24	0.24	0.25	0.19	0.25	0.19	0.22					
	8.0	0.21	0.31	0.26	0.25	0.19	0.21	0.22	0.22					
	12.0	0.17	0.23	0.22	0.20	0.16	0.19	0.17	0.17					
	17.7	0.16	0.19	0.19	0.17	0.12	0.14	0.12	0.15					
37	2.0	0.96	0.96	1.00	1.00	1.00	1.00	1.04	1.00	0.73	0.17	0.47	1.04	0.57
	4.1	0.86	0.90	0.90	0.86	0.92	0.90	0.92	1.02					
	6.0	0.58	0.64	0.50	0.64	0.53	0.58	0.57	0.53					
	8.0	0.61	0.64	0.54	0.54	0.66	0.64	0.61	0.68					
	10.1	0.75	0.71	0.66	0.64	0.68	0.72	0.70	0.83					
	12.0	0.58	0.63	0.54	0.47	0.57	0.58	0.63	0.72					
38	1.4	0.12	0.12	0.12	0.12	0.12	0.18	0.18	0.18	0.15	0.04	0.08	0.23	0.15
	4.0	0.10	0.15	0.15	0.13	0.10	0.08	0.13	0.10					
	6.1	0.12	0.11	0.16	0.11	0.15	0.18	0.16	0.16					
	8.0	0.21	0.21	0.22	0.19	0.21	0.21	0.23	0.22					
	12.0	0.15	0.17	0.19	0.18	0.17	0.17	0.15	0.19					
	17.2	0.13	0.17	0.11	0.13	0.11	0.14	0.14	0.12					
39	1.4	0.60	0.72	0.72	0.78	0.72	0.78	0.72	1.25	0.51	0.17	0.32	1.25	0.94
	4.0	0.65	0.58	0.61	0.56	0.67	0.63	0.67	0.61					
	6.1	0.32	0.44	0.36	0.37	0.44	0.42	0.44	0.44					
	8.0	0.43	0.47	0.48	0.38	0.51	0.50	0.51	0.53					
	9.9	0.39	0.44	0.41	0.38	0.45	0.42	0.40	0.41					
	12.0	0.33	0.38	0.42	0.36	0.44	0.39	0.45	0.35					

Table B3 - Mass Loss Rates for Sites (pg. 6 of 6)

Site No.	Exp Time yrs	Mass Loss Rate (g/m²/d) - 2 Sample Average								Site Average (g/m²/d)				
		Nom. 1.5 in. Pipe				Nom. 3.0 in. Pipe				Mean	Std. Dev.	Min.	Max.	Range
		a	b	e	y	B	K	M	Y					
40	2.0	0.88	0.88	0.88	0.84	0.96	1.04	0.92	1.17	0.74	0.16	0.47	1.17	0.70
	4.1	0.82	0.88	0.79	0.75	0.86	0.88	0.94	0.86					
	6.0	0.75	0.77	0.70	0.74	0.85	0.92	0.85	0.78					
	8.0	0.67	0.68	0.68	0.57	0.73	0.74	0.86	0.80					
	10.0	0.58	0.60	0.61	0.50	0.68	0.60	0.78	0.63					
	12.0	0.48	0.55	0.50	0.47	0.60	0.60	0.49	0.58					
41	1.5	0.45	0.28	0.28	0.56	0.33	0.33	0.22	0.22	0.45	0.13	0.22	0.75	0.53
	4.0	0.56	0.75	0.63	0.63	0.56	0.61	0.50	0.52					
	6.0	0.64	0.65	0.63	0.58	0.45	0.50	0.57	0.54					
	7.9	0.51	0.57	0.53	0.50	0.44	0.44	0.49	0.47					
	12.0	0.42	0.43	0.44	0.37	0.45	0.44	0.42	0.42					
	17.4	0.28	0.36	0.33	0.34	0.26	0.26	0.34	0.34					
42	2.0	1.25	1.50	1.38	1.30	1.21	1.00	1.21	0.79	0.95	0.23	0.58	1.50	0.92
	4.1	0.92	1.49	1.16	1.20	0.98	0.69	1.04	0.69					
	6.0	0.97	1.11	1.07	1.02	0.79	0.96	0.86	0.92					
	8.0	0.62	0.80	0.86	0.97	0.67	0.71	0.70	0.58					
	10.1	0.95	0.88	1.00	1.03	0.73	0.92	0.77	0.76					
	12.0	0.74	1.19	0.87	1.21	0.62	0.78	0.77	0.93					
43	1.3	0.90	1.09	1.22	1.80	1.29	1.41	1.48	1.54	1.12	0.24	0.73	1.80	1.07
	4.1	0.73	1.47	0.94	0.98	0.96	1.06	1.41	1.30					
	6.2	0.78	1.08	1.31	1.00	0.97	0.84	1.04	1.12					
	8.0	0.95	1.60	1.08	1.19	1.05	1.13	0.94	1.50					
	9.9	0.98	0.96	0.89	1.43	0.88	0.95	0.83	1.01					
	12.0	1.08	1.16	1.36	1.23	0.98	0.88	0.91	1.13					
44	1.1	0.23	0.46	0.30	0.38	0.30	0.23	0.30	0.30	0.29	0.07	0.19	0.46	0.28
	3.6	0.32	0.42	0.46	0.42	0.32	0.32	0.28	0.30					
	5.7	0.34	0.32	0.34	0.35	0.31	0.32	0.29	0.29					
	7.6	0.19	0.25	0.24	0.22	0.21	0.23	0.22	0.23					
	11.6	0.21	0.30	0.34	0.25	0.24	0.20	0.24	0.23					
45	1.2	0.84	0.97	0.97	0.91	0.77	0.70	1.18	0.97	0.71	0.25	0.33	1.18	0.85
	3.8	0.79	0.66	0.64	0.73	0.64	0.70	0.62	0.73					
	5.8	0.42	0.46	0.33	0.45	0.48	0.45	0.46	0.43					
	7.7	0.38	0.61	0.46	0.41	0.44	0.42	0.41	0.46					
	9.8	1.17	1.01	1.01	1.03	1.07	1.12	1.05	1.15					
	11.7	0.69	0.64	0.69	0.66	0.76	0.81	0.66	0.80					
46	1.5	0.45	0.72	0.67	0.50	0.56	0.61	0.67	0.67	0.49	0.13	0.28	0.72	0.45
	4.0	0.56	0.67	0.61	0.54	0.50	0.54	0.56	0.67					
	5.1	0.48	0.46	0.51	0.49	0.46	0.49	0.43	0.52					
	8.0	0.58	0.65	0.54	0.60	0.60	0.62	0.61	0.70					
	10.2	0.33	0.38	0.39	0.34	0.36	0.29	0.35	0.32					
	12.0	0.28	0.36	0.31	0.31	0.33	0.30	0.33	0.33					
47	1.5	0.28	0.28	2.78	0.22	0.22	0.33	0.28	0.22	0.31	0.37	0.16	2.78	2.63
	4.1	0.26	0.39	0.35	0.26	0.26	0.26	0.33	0.41					
	6.1	0.16	0.21	0.22	0.16	0.26	0.22	0.30	0.21					
	8.0	0.23	0.22	0.18	0.19	0.22	0.22	0.21	0.17					
	12.1	0.16	0.27	0.21	0.19	0.22	0.21	0.19	0.21					
	17.4	0.28	0.31	0.38	0.40	0.29	0.31	0.26	0.24					

Table B4 - Measured Maximum Pipe Wall Penetrations (pg. 1 of 6)

Site No.	Exp Time yrs	Maximum Penetration (mm) - 2 Sample Average								Site-Retrieval Average (mm)				
		Nom. 1.5 in. Pipe				Nom. 3.0 in. Pipe				Mean	Std. Dev.	Min.	Max.	Range
		abeyBKMY												
1	1.0	0.25	0.25	0.25	0.25	0.25	0.25	0.25	0.25	0.25	0.00	0.25	0.25	0.00
	3.6	1.17	0.84	1.02	1.02	1.07	0.91	1.14	1.27	1.05	0.14	0.84	1.27	0.43
	5.5	1.40	1.14	0.97	1.40	1.47	1.40	1.27	1.24	1.29	0.17	0.97	1.47	0.51
	7.7	1.12	1.88	1.32	1.30	1.35	1.37	1.57	1.14	1.38	0.25	1.12	1.88	0.76
	9.6	1.63	1.85	2.18	2.31	2.44	2.39	2.29	2.49	2.20	0.30	1.63	2.49	0.86
	11.6	2.34	1.83	2.13	1.73	1.93	2.39	2.57	3.18	2.26	0.47	1.73	3.18	1.45
2	2.1	0.25	0.25	0.25	0.25	0.25	0.25	0.25	0.25	0.25	0.00	0.25	0.25	0.00
	4.0	1.40	1.04	0.97	1.19	1.24	1.14	1.12	1.19	1.16	0.13	0.97	1.40	0.43
	5.9	1.07	1.02	1.17	1.04	0.94	1.09	1.07	1.52	1.11	0.18	0.94	1.52	0.58
	7.9	1.22	0.91	1.04	1.27	1.17	1.14	1.27	1.22	1.16	0.12	0.91	1.27	0.36
	12.0	1.42	1.22	1.40	2.03	1.42	1.37	1.24	1.70	1.48	0.27	1.22	2.03	0.81
	17.6	1.80	1.52	1.42	1.63	1.19	1.78	1.47	1.70	1.57	0.20	1.19	1.80	0.61
3	2.0	1.57	1.17	1.12	1.42	1.40	1.35	1.27	1.52	1.35	0.16	1.12	1.57	0.46
	4.1	1.88	1.68	1.98	1.63	1.60	1.35	1.85	2.29	1.78	0.28	1.35	2.29	0.94
	6.0	1.75	1.52	1.78	1.83	1.83	2.08	1.78	1.65	1.78	0.16	1.52	2.08	0.56
	8.0	1.57	1.47	1.78	1.88	1.57	1.73	1.68	1.73	1.68	0.13	1.47	1.88	0.41
	10.1	3.00	1.78	1.83	1.91	1.88	1.57	2.13	2.03	2.02	0.43	1.57	3.00	1.42
	12.1	2.26	2.03	1.57	1.40	1.93	1.73	1.83	2.13	1.86	0.29	1.40	2.26	0.86
4	1.4	0.25	0.25	0.25	0.25	0.25	0.25	0.25	0.25	0.25	0.00	0.25	0.25	0.00
	4.0	0.84	0.69	0.66	0.61	0.66	0.81	0.76	1.07	0.76	0.15	0.61	1.07	0.46
	6.1	2.08	1.19	1.12	1.19	1.27	1.42	1.17	2.29	1.47	0.46	1.12	2.29	1.17
	8.0	1.52	0.91	0.86	1.12	0.97	1.47	1.04	1.93	1.23	0.38	0.86	1.93	1.07
	12.0	3.68	1.98	2.01	2.08	2.21	2.74	2.13	3.86	2.59	0.77	1.98	3.86	1.88
5	1.9	0.25	0.25	0.25	0.25	0.25	0.25	0.25	0.25	0.25	0.00	0.25	0.25	0.00
	4.1	0.74	0.56	0.91	0.81	0.71	0.79	0.76	0.81	0.76	0.10	0.56	0.91	0.36
	6.2	1.12	0.89	1.27	0.86	1.02	1.57	1.57	0.97	1.16	0.29	0.86	1.57	0.71
	8.1	0.91	1.37	1.12	0.97	0.97	1.32	1.47	1.14	1.16	0.21	0.91	1.47	0.56
	12.1	0.84	1.27	1.07	0.94	1.12	1.22	1.17	1.24	1.11	0.15	0.84	1.27	0.43
	17.5	1.93	1.14	1.30	1.07	1.68	2.31	1.57	1.80	1.60	0.42	1.07	2.31	1.24
6	1.9	0.36	0.66	0.66	0.51	0.56	0.46	0.48	0.53	0.53	0.10	0.36	0.66	0.30
	4.1	0.56	0.56	0.53	0.51	0.41	0.56	0.58	0.41	0.51	0.07	0.41	0.58	0.18
	6.2	0.56	0.36	0.41	0.53	0.58	0.64	0.56	0.46	0.51	0.10	0.36	0.64	0.28
	8.1	0.43	0.51	0.30	0.43	0.51	0.56	0.46	0.48	0.46	0.08	0.30	0.56	0.25
	12.1	0.58	0.51	0.46	0.46	0.76	0.41	0.66	0.81	0.58	0.15	0.41	0.81	0.41
	17.5	0.69	0.76	0.46	0.51	0.81	0.76	0.51	0.66	0.64	0.14	0.46	0.81	0.36
7	1.0	0.25	0.25	0.25	0.25	0.25	0.25	0.25	0.25	0.25	0.00	0.25	0.25	0.00
	3.5	0.74	0.51	0.56	0.41	0.46	0.81	0.51	0.48	0.56	0.14	0.41	0.81	0.41
	7.7	0.58	0.86	0.64	0.64	0.91	1.12	0.99	0.56	0.79	0.21	0.56	1.12	0.56
	11.5	1.32	1.02	1.27	0.74	1.19	1.12	1.22	1.22	1.14	0.19	0.74	1.32	0.58
	16.9	1.22	0.91	1.27	1.55	1.88	1.70	1.22	1.42	1.40	0.31	0.91	1.88	0.97
8	1.1	1.12	0.76	0.97	0.76	0.97	1.09	1.19	1.57	1.05	0.26	0.76	1.57	0.81
	3.8	1.17	0.91	1.12	0.94	0.97	1.30	1.22	1.45	1.13	0.19	0.91	1.45	0.53
	5.8	1.57	1.32	1.32	1.37	1.73	1.96	1.42	1.78	1.56	0.24	1.32	1.96	0.64
	7.7	1.57	1.32	1.68	1.32	1.57	2.18	1.91	2.39	1.74	0.39	1.32	2.39	1.07
	9.9	1.88	1.68	1.55	1.70	1.60	2.36	1.83	1.91	1.81	0.26	1.55	2.36	0.81
	11.8	2.54	1.93	1.88	1.47	2.11	2.34	2.79	3.23	2.29	0.56	1.47	3.23	1.75

Table B4 - Measured Maximum Pipe Wall Penetrations (pg. 2 of 6)

Site No.	Exp Time yrs	Maximum Penetration (mm) - 2 Sample Average								Site-Retrieval Average (mm)				
		Nom. 1.5 in. Pipe				Nom. 3.0 in. Pipe				Mean	Std. Dev.	Min.	Max.	Range
		a	b	e	y	B	K	M	Y					
9	1.0	1.14	0.76	0.61	0.56	0.76	0.81	0.74	0.86	0.78	0.18	0.56	1.14	0.58
	3.5	1.17	1.27	1.37	1.37	1.37	1.47	1.73	1.63	1.42	0.18	1.17	1.73	0.56
	5.5	0.46	0.79	0.71	0.69	0.81	1.14	0.91	0.81	0.79	0.20	0.46	1.14	0.69
	7.7	0.91	1.17	0.97	0.81	1.09	1.22	1.17	1.02	1.04	0.14	0.81	1.22	0.41
	11.5	1.07	1.07	0.81	1.32	1.02	1.22	1.24	2.44	1.27	0.50	0.81	2.44	1.63
	16.9	1.75	1.30	1.63	1.65	1.73	1.42	1.57	2.77	1.73	0.45	1.30	2.77	1.47
10	1.3	0.66	0.38	0.46	0.41	0.74	0.51	0.56	0.64	0.54	0.13	0.38	0.74	0.36
	4.0	0.51	0.51	0.51	0.51	0.41	0.51	0.38	0.51	0.48	0.05	0.38	0.51	0.13
	6.1	1.17	0.97	0.99	1.07	1.22	0.97	1.04	1.27	1.09	0.12	0.97	1.27	0.30
	7.9	1.22	0.81	0.97	0.94	1.22	1.42	1.02	1.07	1.08	0.19	0.81	1.42	0.61
	12.0	1.27	1.32	1.02	0.97	1.14	1.32	1.37	1.68	1.26	0.22	0.97	1.68	0.71
11	1.4	1.24	0.97	0.81	0.71	0.97	0.84	0.97	1.07	0.95	0.16	0.71	1.24	0.53
	4.0	1.57	1.70	1.27	1.55	1.73	1.37	1.98	1.57	1.59	0.22	1.27	1.98	0.71
	6.0	1.73	1.80	1.55	1.78	1.57	1.55	2.34	1.85	1.77	0.26	1.55	2.34	0.79
	7.8	1.63	1.63	1.42	1.17	1.68	1.47	1.83	1.57	1.55	0.20	1.17	1.83	0.66
	10.0	1.91	1.52	1.93	1.78	2.29	1.68	1.68	1.80	1.82	0.23	1.52	2.29	0.76
	11.9	2.51	1.91	1.80	1.47	2.08	1.78	2.03	2.24	1.98	0.32	1.47	2.51	1.04
12	1.9	0.25	0.25	0.25	0.25	0.25	0.25	0.25	0.25	0.25	0.00	0.25	0.25	0.00
	4.1	1.12	1.09	1.12	1.07	1.17	1.02	1.07	1.65	1.16	0.20	1.02	1.65	0.64
	6.2	1.37	1.52	1.65	1.42	1.73	1.37	1.60	1.52	1.52	0.13	1.37	1.73	0.36
	8.0	0.66	0.43	0.66	0.51	0.61	0.46	0.46	0.64	0.55	0.10	0.43	0.66	0.23
	12.1	1.52	1.47	1.40	1.50	1.88	1.32	1.65	2.16	1.61	0.28	1.32	2.16	0.84
	17.5	1.78	1.22	1.83	1.63	1.93	1.42	2.18	1.75	1.72	0.30	1.22	2.18	0.97
13	1.9	0.91	0.81	1.22	0.91	1.19	1.65	1.60	1.80	1.26	0.38	0.81	1.80	0.99
	4.2	0.69	1.40	1.19	0.94	0.97	1.47	1.57	1.63	1.23	0.34	0.69	1.63	0.94
	5.9	1.24	2.46	1.70	2.16	1.50	1.70	1.91	1.40	1.76	0.40	1.24	2.46	1.22
14	1.1	0.86	0.97	0.66	0.71	0.81	0.81	1.02	1.65	0.94	0.31	0.66	1.65	0.99
	3.8	1.37	1.52	1.32	1.35	1.60	1.42	1.63	1.88	1.51	0.19	1.32	1.88	0.56
	5.8	2.21	2.08	2.13	2.01	2.49	2.29	2.39	3.89	2.44	0.61	2.01	3.89	1.88
	7.7	2.08	2.18	2.72	2.62	2.26	1.96	3.43	4.09	2.67	0.74	1.96	4.09	2.13
	9.9	3.05	2.69	3.30	3.33	2.46	2.46	2.90	3.61	2.97	0.42	2.46	3.61	1.14
	11.8	2.57	2.77	2.03	2.13	3.23	2.11	2.39	2.84	2.51	0.42	2.03	3.23	1.19
15	2.0	0.76	0.86	1.02	0.89	1.02	0.86	1.12	1.12	0.96	0.13	0.76	1.12	0.36
	4.0	1.42	0.66	0.89	0.71	0.76	0.86	1.09	1.22	0.95	0.27	0.66	1.42	0.76
	5.9	1.65	1.50	1.37	1.35	1.63	1.63	1.63	1.47	1.53	0.12	1.35	1.65	0.30
	8.0	1.50	1.68	1.57	0.97	1.65	1.14	1.12	1.32	1.37	0.27	0.97	1.68	0.71
	12.0	1.98	1.27	1.42	1.37	1.60	2.08	1.83	1.57	1.64	0.30	1.27	2.08	0.81
	17.6	1.60	1.45	1.45	1.47	1.55	1.68	1.32	1.57	1.51	0.11	1.32	1.68	0.36
16	2.0	1.09	0.91	0.97	1.02	1.07	0.89	0.99	1.14	1.01	0.09	0.89	1.14	0.25
	4.0	1.17	1.17	1.37	1.19	1.17	1.17	1.83	1.78	1.36	0.29	1.17	1.83	0.66
	6.0	2.34	2.13	2.39	3.05	2.13	1.57	2.44	2.29	2.29	0.41	1.57	3.05	1.47
	7.9	1.50	2.13	1.73	1.70	1.75	2.13	1.75	1.88	1.82	0.22	1.50	2.13	0.64
	10.0	1.75	1.55	1.63	1.42	1.88	1.78	1.68	1.80	1.69	0.15	1.42	1.88	0.46
	12.0	2.08	1.55	1.47	1.57	1.42	2.18	1.78	2.13	1.77	0.32	1.42	2.18	0.76

Table B4 - Measured Maximum Pipe Wall Penetrations (pg. 3 of 6)

Site No.	Exp Time yrs	Maximum Penetration (mm) - 2 Sample Average								Site-Retrieval Average (mm)				
		Nom. 1.5 in. Pipe				Nom. 3.0 in. Pipe				Mean	Std. Dev.	Min.	Max.	Range
		a	b	e	y	B	K	M	Y					
17	1.2	0.25	0.25	0.25	0.25	0.25	0.25	0.25	0.25	0.25	0.00	0.25	0.25	0.00
	3.8	0.56	0.41	0.56	0.51	0.58	0.61	0.81	0.53	0.57	0.12	0.41	0.81	0.41
	5.9	0.81	0.66	0.66	0.81	0.81	0.81	0.79	0.89	0.78	0.08	0.66	0.89	0.23
	7.7	0.81	0.91	0.91	0.76	1.02	0.99	0.91	0.97	0.91	0.09	0.76	1.02	0.25
	11.8	1.04	0.97	1.07	0.99	1.07	1.19	1.22	1.07	1.08	0.09	0.97	1.22	0.25
	17.0	1.07	0.97	1.04	0.86	1.09	1.27	1.22	1.45	1.12	0.18	0.86	1.45	0.58
18	1.2	0.69	0.51	0.48	0.36	0.61	0.71	0.41	0.71	0.56	0.14	0.36	0.71	0.36
	3.8	1.02	1.02	1.22	0.86	0.86	1.32	1.09	0.94	1.04	0.16	0.86	1.32	0.46
	5.8	1.80	1.83	1.80	1.70	1.63	1.68	1.57	2.03	1.76	0.14	1.57	2.03	0.46
	7.7	1.40	0.94	1.02	1.07	1.42	1.45	1.07	1.27	1.20	0.20	0.94	1.45	0.51
	9.8	1.17	1.17	1.30	1.45	1.37	1.63	1.27	1.17	1.31	0.16	1.17	1.63	0.46
	11.7	1.32	1.07	1.04	0.97	1.04	1.78	1.12	1.12	1.18	0.26	0.97	1.78	0.81
19	1.1	0.97	0.61	0.66	0.61	0.71	0.64	0.81	0.97	0.75	0.15	0.61	0.97	0.36
	3.7	0.97	1.14	1.27	0.97	0.91	1.17	1.50	1.27	1.15	0.20	0.91	1.50	0.58
	5.7	1.17	1.27	1.22	1.22	1.47	1.50	1.47	1.27	1.32	0.13	1.17	1.50	0.33
	7.6	1.12	1.02	1.17	1.02	1.42	1.47	1.42	1.68	1.29	0.24	1.02	1.68	0.66
	9.7	1.57	1.30	1.55	1.40	1.42	1.98	1.65	1.73	1.57	0.22	1.30	1.98	0.69
	11.6	1.42	1.57	1.80	1.68	1.68	2.16	1.52	1.60	1.68	0.22	1.42	2.16	0.74
20	1.0	0.25	0.25	0.25	0.25	0.25	0.25	0.25	0.25	0.25	0.00	0.25	0.25	0.00
	3.6	0.51	0.51	0.51	0.51	0.51	0.51	0.51	0.51	0.51	0.00	0.51	0.51	0.00
	5.5	0.51	0.66	0.56	0.81	0.74	0.61	0.86	0.71	0.68	0.12	0.51	0.86	0.36
	7.7	0.71	0.97	1.02	1.07	0.81	0.89	1.07	1.65	1.02	0.28	0.71	1.65	0.94
	9.6	0.81	1.32	0.97	0.56	1.12	0.91	1.07	0.91	0.96	0.22	0.56	1.32	0.76
	11.6	1.70	1.22	1.83	1.63	1.14	2.03	1.42	1.57	1.57	0.30	1.14	2.03	0.89
21	1.5	0.25	0.25	0.25	0.25	0.25	0.25	0.25	0.25	0.25	0.00	0.25	0.25	0.00
	4.0	1.55	1.02	1.27	1.60	1.52	0.99	1.22	1.12	1.29	0.24	0.99	1.60	0.61
	6.0	1.80	1.32	1.52	1.47	1.52	1.50	1.68	1.52	1.54	0.14	1.32	1.80	0.48
22	1.7	1.12	0.89	1.12	1.12	0.97	0.99	1.32	1.19	1.09	0.14	0.89	1.32	0.43
	3.7	1.14	1.17	1.30	1.12	1.47	1.52	1.37	1.27	1.30	0.15	1.12	1.52	0.41
	5.6	1.73	1.40	1.47	1.37	1.57	1.57	1.57	1.52	1.53	0.11	1.37	1.73	0.36
	7.6	1.09	1.42	1.60	1.32	1.42	1.47	1.68	1.75	1.47	0.21	1.09	1.75	0.66
	9.6	1.65	1.57	1.37	1.45	1.73	1.65	1.60	1.73	1.59	0.13	1.37	1.73	0.36
	11.6	1.83	1.68	1.68	1.98	1.52	1.42	1.65	1.80	1.70	0.18	1.42	1.98	0.56
23	1.9	1.35	1.22	1.22	1.27	1.22	2.18	1.32	1.70	1.44	0.34	1.22	2.18	0.97
	4.3	2.39	2.62	2.90	2.11	2.49	3.40	2.74	2.46	2.64	0.39	2.11	3.40	1.30
	6.2	3.68	3.15	2.29	2.54	2.31	2.77	2.54	3.30	2.82	0.51	2.29	3.68	1.40
	8.0	3.68	3.68	3.68	3.25	4.01	4.04	3.40	5.49	3.91	0.69	3.25	5.49	2.24
	10.2	3.68	3.68	3.68	3.68	3.45	3.84	3.45	4.14	3.70	0.22	3.45	4.14	0.69
	12.1	3.68	3.68	3.68	3.68	3.81	3.81	4.14	4.50	3.87	0.30	3.68	4.50	0.81
24	1.3	0.25	0.25	0.25	0.25	0.25	0.25	0.25	0.25	0.25	0.00	0.25	0.25	0.00
	4.0	0.51	0.51	0.51	0.51	0.51	0.51	0.51	0.51	0.51	0.00	0.51	0.51	0.00
	6.1	0.51	0.51	0.51	0.51	0.53	0.51	0.51	0.71	0.54	0.07	0.51	0.71	0.20
	7.9	0.56	0.51	0.51	0.51	0.48	0.64	0.51	0.79	0.56	0.10	0.48	0.79	0.30
	12.0	0.71	0.58	0.53	0.64	0.71	0.71	0.76	0.79	0.68	0.09	0.53	0.79	0.25
	17.2	0.66	0.61	0.41	0.66	0.76	0.69	0.91	0.71	0.68	0.14	0.41	0.91	0.51

Table B4 - Measured Maximum Pipe Wall Penetrations (pg. 4 of 6)

Site No.	Exp Time yrs	Maximum Penetration (mm) - 2 Sample Average								Site-Retrieval Average (mm)				
		Nom. 1.5 in. Pipe				Nom. 3.0 in. Pipe				Mean	Std. Dev.	Min.	Max.	Range
		a	b	e	y	B	K	M	Y					
25	1.0	0.61	0.48	0.46	0.64	0.56	0.69	0.71	0.64	0.60	0.09	0.46	0.71	0.25
	3.7	0.86	0.76	0.69	0.76	0.79	0.74	0.97	0.86	0.80	0.09	0.69	0.97	0.28
	5.7	1.22	1.04	0.99	0.91	1.19	1.22	1.47	1.32	1.17	0.18	0.91	1.47	0.56
	7.6	1.07	0.89	0.91	0.86	1.32	0.97	1.19	1.07	1.04	0.16	0.86	1.32	0.46
	11.7	1.32	1.09	1.22	0.81	1.22	1.17	1.45	1.32	1.20	0.19	0.81	1.45	0.64
	17.0	1.91	1.27	1.07	1.07	1.37	1.57	1.42	1.45	1.39	0.27	1.07	1.91	0.84
26	1.0	0.25	0.25	0.25	0.25	0.25	0.25	0.25	0.25	0.25	0.00	0.25	0.25	0.00
	3.5	0.99	1.14	1.07	1.17	0.91	1.14	1.68	1.22	1.17	0.23	0.91	1.68	0.76
	5.5	1.78	1.68	1.70	1.63	1.83	1.68	1.98	2.03	1.79	0.15	1.63	2.03	0.41
	7.7	0.99	1.19	0.97	1.12	1.19	1.22	1.22	1.47	1.17	0.16	0.97	1.47	0.51
	11.5	0.97	1.19	1.12	0.89	1.12	1.17	0.97	1.09	1.06	0.11	0.89	1.19	0.30
	16.9	1.09	1.07	1.07	0.91	1.07	1.52	1.47	1.35	1.19	0.22	0.91	1.52	0.61
27	2.0	0.25	0.25	0.25	0.25	0.25	0.25	0.25	0.25	0.25	0.00	0.25	0.25	0.00
	4.0	0.51	0.56	0.79	0.84	0.91	0.97	0.69	1.07	0.79	0.20	0.51	1.07	0.56
	6.0	0.53	0.71	0.81	1.02	0.76	1.42	0.89	1.17	0.91	0.28	0.53	1.42	0.89
	8.0	0.79	0.86	0.66	0.91	1.04	1.83	1.17	1.07	1.04	0.36	0.66	1.83	1.17
	12.0	1.07	1.22	1.52	1.52	1.40	2.03	2.08	1.88	1.59	0.37	1.07	2.08	1.02
	17.6	0.99	1.47	1.75	1.37	1.88	2.34	2.13	1.98	1.74	0.44	0.99	2.34	1.35
28	1.6	0.51	0.66	0.86	0.66	0.74	1.02	1.40	0.91	0.84	0.28	0.51	1.40	0.89
	5.6	1.27	1.75	1.52	1.40	1.42	1.93	1.42	1.91	1.58	0.25	1.27	1.93	0.66
	7.7	3.48	3.15	2.97	2.69	3.35	4.09	2.97	3.89	3.32	0.48	2.69	4.09	1.40
	9.6	3.68	3.35	3.48	3.68	4.24	4.65	3.86	5.49	4.05	0.71	3.35	5.49	2.13
29	2.0	1.22	0.94	1.37	1.42	1.07	1.73	1.09	0.94	1.22	0.27	0.94	1.73	0.79
	4.1	1.65	1.19	1.52	1.73	1.83	1.98	1.52	2.29	1.71	0.33	1.19	2.29	1.09
	6.0	2.64	1.98	2.34	2.11	1.42	2.03	2.29	2.06	2.11	0.35	1.42	2.64	1.22
	8.0	2.64	1.73	2.29	2.29	2.49	2.29	3.25	4.52	2.69	0.86	1.73	4.52	2.79
	10.0	3.68	2.46	1.88	2.34	2.13	3.05	2.54	4.52	2.83	0.89	1.88	4.52	2.64
	12.0	3.68	2.29	3.45	3.68	3.40	5.49	3.02	5.49	3.81	1.13	2.29	5.49	3.20
30	1.1	0.61	0.25	0.30	0.25	0.30	0.41	0.41	0.46	0.37	0.12	0.25	0.61	0.36
	3.6	0.51	0.51	0.51	0.51	0.51	0.51	0.51	0.51	0.51	0.00	0.51	0.51	0.00
	5.7	0.51	0.51	0.61	0.41	0.71	0.71	0.69	0.71	0.61	0.12	0.41	0.71	0.30
	8.2	1.17	0.91	0.84	0.86	1.57	0.89	1.22	0.79	1.03	0.27	0.79	1.57	0.79
	11.6	1.37	1.30	1.47	1.30	1.35	1.37	1.60	1.68	1.43	0.14	1.30	1.68	0.38
	17.0	1.27	1.12	1.32	1.07	1.35	1.63	1.93	1.65	1.42	0.30	1.07	1.93	0.86
31	2.0	0.76	0.84	1.70	0.99	0.94	1.22	1.32	1.37	1.14	0.32	0.76	1.70	0.94
	4.1	0.61	0.66	0.64	0.64	0.76	1.19	0.71	0.86	0.76	0.19	0.61	1.19	0.58
	6.0	0.71	0.66	0.76	1.35	0.74	1.42	0.81	1.17	0.95	0.31	0.66	1.42	0.76
	8.0	0.91	1.12	0.94	0.97	1.07	1.14	1.63	0.91	1.09	0.24	0.91	1.63	0.71
	12.0	0.86	0.74	0.86	0.76	1.04	1.04	1.02	0.91	0.90	0.12	0.74	1.04	0.30
	17.7	1.27	1.07	1.09	1.32	1.04	2.29	1.68	1.24	1.37	0.42	1.04	2.29	1.24
32	1.0	1.02	0.66	0.53	0.56	0.76	0.58	0.69	0.66	0.68	0.15	0.53	1.02	0.48
	3.7	1.12	1.02	0.76	0.69	1.02	0.86	0.91	1.04	0.93	0.15	0.69	1.12	0.43
	5.8	0.94	0.81	0.76	0.91	0.97	0.99	0.94	1.07	0.92	0.10	0.76	1.07	0.30
	7.6	0.81	1.07	1.07	0.99	0.97	1.22	1.12	1.17	1.05	0.13	0.81	1.22	0.41
	9.6	1.09	1.40	1.17	1.12	1.32	1.24	1.37	2.39	1.39	0.42	1.09	2.39	1.30
	11.7	1.47	1.07	1.17	1.27	1.50	2.18	1.57	2.36	1.57	0.47	1.07	2.36	1.30

Table B4 - Measured Maximum Pipe Wall Penetrations (pg. 5 of 6)

Site No.	Exp Time yrs	Maximum Penetration (mm) - 2 Sample Average								Site-Retrieval Average (mm)				
		Nom. 1.5 in. Pipe				Nom. 3.0 in. Pipe				Mean	Std. Dev.	Min.	Max.	Range
		a	b	e	y	B	K	M	Y					
33	1.0	0.25	0.25	0.25	0.25	0.25	0.25	0.25	0.25	0.25	0.00	0.25	0.25	0.00
	3.7	0.97	0.51	0.51	0.64	0.81	0.76	1.02	0.97	0.77	0.20	0.51	1.02	0.51
	5.8	1.17	0.91	0.51	1.30	0.91	1.37	0.84	0.71	0.97	0.30	0.51	1.37	0.86
	7.6	3.30	1.52	1.68	2.46	2.54	2.97	2.44	2.39	2.41	0.59	1.52	3.30	1.78
	9.7	2.90	2.08	2.34	1.88	2.84	2.59	2.92	2.82	2.55	0.40	1.88	2.92	1.04
	11.7	2.84	2.49	2.29	2.64	2.64	2.74	2.44	2.13	2.53	0.24	2.13	2.84	0.71
34	1.4	0.51	0.53	0.51	0.46	0.56	0.69	0.56	0.74	0.57	0.09	0.46	0.74	0.28
	4.0	0.58	0.53	0.51	0.46	0.41	0.46	0.41	0.43	0.47	0.06	0.41	0.58	0.18
	6.1	1.22	0.71	0.41	0.81	0.91	0.97	0.74	1.22	0.87	0.27	0.41	1.22	0.81
	8.0	1.07	1.17	1.14	1.32	1.09	1.70	1.02	1.73	1.28	0.28	1.02	1.73	0.71
	9.9	2.08	1.22	2.13	2.39	1.80	1.85	1.96	2.64	2.01	0.42	1.22	2.64	1.42
	12.0	1.47	0.91	1.04	0.81	1.40	0.91	1.27	1.88	1.21	0.36	0.81	1.88	1.07
35	1.9	0.25	0.43	0.25	0.25	0.25	0.25	0.43	0.46	0.32	0.10	0.25	0.46	0.20
	4.1	0.41	0.51	0.51	0.51	0.51	0.43	0.56	0.61	0.50	0.06	0.41	0.61	0.20
	6.2	0.51	0.51	1.02	0.51	0.56	0.51	0.46	0.53	0.57	0.18	0.46	1.02	0.56
	8.0	0.81	0.61	0.58	0.51	0.46	0.97	0.71	2.46	0.89	0.66	0.46	2.46	2.01
	12.1	0.25	0.61	0.46	0.51	0.91	0.56	1.75	0.69	0.72	0.46	0.25	1.75	1.50
	17.5	0.30	1.37	0.23	0.43	0.81	0.86	0.91	1.63	0.82	0.50	0.23	1.63	1.40
36	2.0	0.71	0.94	0.86	0.97	0.94	1.22	1.09	0.94	0.96	0.15	0.71	1.22	0.51
	4.1	0.94	0.97	0.76	0.79	0.99	1.27	1.17	1.07	0.99	0.17	0.76	1.27	0.51
	6.0	1.19	1.22	0.94	1.04	1.04	1.32	1.22	1.07	1.13	0.13	0.94	1.32	0.38
	8.0	1.22	1.37	0.89	0.97	1.22	1.47	1.27	1.27	1.21	0.19	0.89	1.47	0.58
	12.0	1.42	1.17	1.22	1.22	1.19	1.52	1.14	1.12	1.25	0.14	1.12	1.52	0.41
	17.7	1.30	1.27	1.40	1.14	1.27	1.52	1.24	1.45	1.32	0.12	1.14	1.52	0.38
37	2.0	0.97	0.89	1.22	1.22	1.07	1.09	1.35	2.29	1.26	0.44	0.89	2.29	1.40
	4.1	1.17	1.07	1.57	1.68	1.50	1.47	1.40	2.34	1.52	0.39	1.07	2.34	1.27
	6.0	1.17	1.30	0.97	1.85	0.97	1.22	1.42	1.50	1.30	0.30	0.97	1.85	0.89
	8.0	1.47	1.24	1.32	1.37	1.63	1.73	1.70	2.44	1.61	0.38	1.24	2.44	1.19
	10.1	1.75	1.80	2.26	1.88	1.91	1.83	2.13	2.90	2.06	0.38	1.75	2.90	1.14
	12.0	1.93	1.63	1.98	1.73	2.03	1.60	2.41	3.23	2.07	0.54	1.60	3.23	1.63
38	1.4	0.25	0.25	0.25	0.25	0.25	0.25	0.25	0.25	0.25	0.00	0.25	0.25	0.00
	4.0	0.51	0.71	0.51	0.51	0.51	0.51	0.51	0.51	0.53	0.07	0.51	0.71	0.20
	6.1	1.02	0.66	0.51	0.51	0.94	0.43	0.61	0.89	0.70	0.22	0.43	1.02	0.58
	8.0	1.32	0.71	0.51	0.61	0.86	0.66	0.74	0.61	0.75	0.25	0.51	1.32	0.81
	12.0	0.89	0.86	0.66	0.51	0.71	0.56	1.07	0.66	0.74	0.19	0.51	1.07	0.56
	17.2	1.02	0.84	0.71	0.91	0.86	0.97	0.86	0.84	0.88	0.09	0.71	1.02	0.30
39	1.4	0.25	0.25	0.25	0.25	0.25	0.25	0.25	0.25	0.25	0.00	0.25	0.25	0.00
	4.0	1.07	0.61	0.97	0.64	0.41	0.61	0.81	1.17	0.78	0.26	0.41	1.17	0.76
	6.1	0.79	1.12	0.97	1.27	1.22	0.97	1.12	1.27	1.09	0.17	0.79	1.27	0.48
	8.0	1.22	1.27	1.17	1.07	1.68	1.83	1.32	1.96	1.44	0.33	1.07	1.96	0.89
	9.9	1.73	1.07	1.07	1.30	1.35	1.73	1.32	2.24	1.47	0.40	1.07	2.24	1.17
	12.0	1.96	1.42	1.27	1.52	1.75	1.75	2.39	2.69	1.84	0.49	1.27	2.69	1.42
40	2.0	1.17	0.71	0.97	0.79	0.71	1.04	0.81	1.07	0.91	0.18	0.71	1.17	0.46
	4.1	1.55	1.32	1.09	0.89	1.04	1.17	1.12	1.17	1.17	0.20	0.89	1.55	0.66
	6.0	1.55	1.75	1.78	1.52	1.65	2.24	1.88	2.13	1.81	0.26	1.52	2.24	0.71
	8.0	2.18	2.57	1.63	1.88	1.78	2.51	2.44	2.34	2.17	0.36	1.63	2.57	0.94
	10.0	1.91	1.57	2.21	1.57	1.65	1.83	1.63	2.18	1.82	0.26	1.57	2.21	0.64
	12.0	3.53	1.73	1.75	2.08	1.75	2.31	1.98	1.91	2.13	0.60	1.73	3.53	1.80

Table B4 - Measured Maximum Pipe Wall Penetrations (pg. 6 of 6)

Site No.	Exp Time yrs	Maximum Penetration (mm) - 2 Sample Average								Site-Retrieval Average (mm)				
		Nom. 1.5 in. Pipe				Nom. 3.0 in. Pipe				Mean	Std. Dev.	Min.	Max.	Range
		a	b	e	y	B	K	M	Y					
41	1.5	0.48	0.99	0.84	1.02	1.12	0.76	1.07	0.81	0.89	0.21	0.48	1.12	0.64
	4.0	0.86	0.91	1.02	1.14	1.19	0.97	1.09	1.17	1.04	0.12	0.86	1.19	0.33
	6.0	1.35	1.27	1.22	1.52	1.40	1.32	1.70	1.32	1.39	0.16	1.22	1.70	0.48
	7.9	1.55	1.65	1.40	1.52	1.47	1.47	1.57	1.42	1.51	0.08	1.40	1.65	0.25
	12.0	2.57	2.39	2.01	2.31	2.18	1.83	2.16	2.03	2.18	0.23	1.83	2.57	0.74
	17.4	3.10	2.39	2.34	2.57	2.18	1.70	2.57	1.98	2.35	0.42	1.70	3.10	1.40
42	2.0	1.35	1.47	1.50	1.27	1.37	1.57	1.98	1.40	1.49	0.22	1.27	1.98	0.71
	4.1	1.63	1.88	1.85	1.93	2.01	2.16	2.08	1.88	1.93	0.16	1.63	2.16	0.53
	6.0	1.96	2.34	2.13	1.88	1.98	2.18	2.41	2.06	2.12	0.19	1.88	2.41	0.53
	8.0	1.93	2.26	2.87	2.82	2.03	2.36	2.24	2.29	2.35	0.34	1.93	2.87	0.94
	10.1	2.13	2.13	2.64	2.34	2.44	3.28	2.62	2.64	2.53	0.37	2.13	3.28	1.14
	12.0	2.39	2.26	2.82	2.18	2.24	3.18	2.34	2.95	2.54	0.38	2.18	3.18	0.99
43	1.3	0.46	0.91	0.61	0.71	1.12	1.14	1.68	1.19	0.98	0.39	0.46	1.68	1.22
	4.1	0.99	1.12	0.89	1.17	2.57	2.01	2.11	1.50	1.54	0.61	0.89	2.57	1.68
	6.2	2.24	2.08	1.70	1.93	2.64	3.35	3.02	1.35	2.29	0.68	1.35	3.35	2.01
	8.0	2.13	2.59	1.93	1.78	2.64	3.45	1.98	2.29	2.35	0.54	1.78	3.45	1.68
	9.9	2.39	1.93	1.78	1.85	3.45	2.54	2.95	3.94	2.60	0.79	1.78	3.94	2.16
	12.0	2.29	2.03	2.54	2.67	3.51	1.98	1.88	1.83	2.34	0.56	1.83	3.51	1.68
44	1.1	0.97	0.91	0.71	0.81	0.66	0.99	0.81	0.97	0.85	0.12	0.66	0.99	0.33
	3.6	1.98	1.09	1.37	1.40	1.17	1.12	1.12	1.27	1.31	0.29	1.09	1.98	0.89
	5.7	1.78	1.32	1.30	1.68	1.42	1.83	1.52	1.73	1.57	0.21	1.30	1.83	0.53
	7.6	1.83	1.24	1.57	1.27	1.42	1.27	1.57	2.24	1.55	0.34	1.24	2.24	0.99
	11.6	2.21	1.42	1.60	1.75	1.65	1.47	2.08	1.88	1.76	0.28	1.42	2.21	0.79
45	1.2	0.25	0.51	0.38	0.33	0.25	0.43	0.76	0.71	0.45	0.19	0.25	0.76	0.51
	3.8	0.91	0.71	0.61	0.61	0.81	0.91	0.79	1.24	0.83	0.21	0.61	1.24	0.64
	5.8	1.14	1.09	1.02	0.86	1.19	0.91	1.22	1.32	1.10	0.16	0.86	1.32	0.46
	7.7	1.27	1.17	1.52	1.14	1.42	1.52	1.52	1.68	1.41	0.19	1.14	1.68	0.53
	9.8	3.63	2.90	3.51	2.97	3.00	3.51	3.25	4.01	3.35	0.39	2.90	4.01	1.12
	11.7	2.08	1.98	2.13	2.08	2.16	2.84	2.49	3.15	2.37	0.42	1.98	3.15	1.17
46	1.5	1.45	1.37	1.40	1.37	1.27	1.02	1.42	2.01	1.41	0.28	1.02	2.01	0.99
	4.0	2.03	1.63	2.01	2.08	1.68	1.47	2.69	2.79	2.05	0.48	1.47	2.79	1.32
	5.1	1.73	1.65	1.68	1.37	1.73	1.17	2.44	2.44	1.77	0.45	1.17	2.44	1.27
	8.0	1.52	2.03	2.74	3.00	1.75	1.73	3.45	3.40	2.45	0.79	1.52	3.45	1.93
	10.2	1.88	2.41	1.73	2.11	2.08	1.68	2.13	2.03	2.01	0.24	1.68	2.41	0.74
	12.0	1.22	1.57	1.63	2.64	1.96	1.57	2.90	2.03	1.94	0.57	1.22	2.90	1.68
47	1.5	0.25	0.25	0.25	0.25	0.25	0.25	0.25	0.25	0.25	0.00	0.25	0.25	0.00
	4.1	0.51	0.51	0.51	0.51	0.51	0.51	0.51	0.51	0.51	0.00	0.51	0.51	0.00
	6.1	0.51	0.51	0.51	0.51	0.51	0.51	0.51	0.51	0.51	0.00	0.51	0.51	0.00
	8.0	0.51	0.51	0.51	0.51	0.51	0.51	0.51	0.51	0.51	0.00	0.51	0.51	0.00
	12.1	0.41	1.17	0.66	0.79	0.86	0.61	0.69	0.66	0.73	0.22	0.41	1.17	0.76
	17.4	1.07	1.35	0.94	1.45	1.30	1.02	1.22	1.17	1.19	0.17	0.94	1.45	0.51

Table B5 - Maximum Penetration Rates (pg. 1 of 6)

Site No.	Exp Time yrs	Maximum Penetration Rate (mm/yr) - 2 Sample Average								Total Site Average (mm/yr)				
		Nom. 1.5 in. Pipe				Nom. 3.0 in. Pipe				Mean	Std. Dev.	Min.	Max.	Range
		a	b	e	y	B	K	M	Y					
1	1.0	0.25	0.25	0.25	0.25	0.25	0.25	0.25	0.25	0.23	0.05	0.15	0.35	0.21
	3.6	0.32	0.23	0.28	0.28	0.30	0.25	0.32	0.35					
	5.5	0.25	0.21	0.18	0.25	0.27	0.25	0.23	0.23					
	7.7	0.15	0.24	0.17	0.17	0.17	0.18	0.20	0.15					
	9.6	0.17	0.19	0.23	0.24	0.25	0.25	0.24	0.26					
	11.6	0.20	0.16	0.18	0.15	0.17	0.21	0.22	0.27					
2	2.1	0.12	0.12	0.12	0.12	0.12	0.12	0.12	0.12	0.16	0.07	0.07	0.35	0.28
	4.0	0.35	0.26	0.24	0.30	0.31	0.29	0.28	0.30					
	5.9	0.18	0.17	0.20	0.18	0.16	0.19	0.18	0.26					
	7.9	0.15	0.12	0.13	0.16	0.15	0.14	0.16	0.15					
	12.0	0.12	0.10	0.12	0.17	0.12	0.11	0.10	0.14					
	17.6	0.10	0.09	0.08	0.09	0.07	0.10	0.08	0.10					
3	2.0	0.79	0.58	0.56	0.71	0.70	0.67	0.64	0.76	0.33	0.19	0.12	0.79	0.67
	4.1	0.46	0.41	0.48	0.40	0.39	0.33	0.45	0.56					
	6.0	0.29	0.25	0.30	0.30	0.30	0.35	0.30	0.28					
	8.0	0.20	0.18	0.22	0.23	0.20	0.22	0.21	0.22					
	10.1	0.30	0.18	0.18	0.19	0.19	0.16	0.21	0.20					
	12.1	0.19	0.17	0.13	0.12	0.16	0.14	0.15	0.18					
4	1.4	0.18	0.18	0.18	0.18	0.18	0.18	0.18	0.18	0.20	0.06	0.11	0.37	0.27
	4.0	0.21	0.17	0.17	0.15	0.17	0.20	0.19	0.27					
	6.1	0.34	0.20	0.18	0.20	0.21	0.23	0.19	0.37					
	8.0	0.19	0.11	0.11	0.14	0.12	0.18	0.13	0.24					
	12.0	0.31	0.17	0.17	0.17	0.18	0.23	0.18	0.32					
5	1.9	0.13	0.13	0.13	0.13	0.13	0.13	0.13	0.13	0.14	0.05	0.06	0.25	0.19
	4.1	0.18	0.14	0.22	0.20	0.17	0.19	0.19	0.20					
	6.2	0.18	0.14	0.20	0.14	0.16	0.25	0.25	0.16					
	8.1	0.11	0.17	0.14	0.12	0.12	0.16	0.18	0.14					
	12.1	0.07	0.10	0.09	0.08	0.09	0.10	0.10	0.10					
	17.5	0.11	0.07	0.07	0.06	0.10	0.13	0.09	0.10					
6	1.9	0.19	0.35	0.35	0.27	0.29	0.24	0.25	0.28	0.10	0.09	0.03	0.35	0.32
	4.1	0.14	0.14	0.13	0.12	0.10	0.14	0.14	0.10					
	6.2	0.09	0.06	0.07	0.09	0.09	0.10	0.09	0.07					
	8.1	0.05	0.06	0.04	0.05	0.06	0.07	0.06	0.06					
	12.1	0.05	0.04	0.04	0.04	0.06	0.03	0.05	0.07					
	17.5	0.04	0.04	0.03	0.03	0.05	0.04	0.03	0.04					
7	1.0	0.25	0.25	0.25	0.25	0.25	0.25	0.25	0.25	0.14	0.07	0.05	0.25	0.20
	3.5	0.21	0.15	0.16	0.12	0.13	0.23	0.15	0.14					
	7.7	0.08	0.11	0.08	0.08	0.12	0.15	0.13	0.07					
	11.5	0.11	0.09	0.11	0.06	0.10	0.10	0.11	0.11					
	16.9	0.07	0.05	0.08	0.09	0.11	0.10	0.07	0.08					
8	1.1	1.02	0.69	0.88	0.69	0.88	0.99	1.09	1.43	0.35	0.29	0.12	1.43	1.31
	3.8	0.31	0.24	0.29	0.25	0.25	0.34	0.32	0.38					
	5.8	0.27	0.23	0.23	0.24	0.30	0.34	0.25	0.31					
	7.7	0.20	0.17	0.22	0.17	0.20	0.28	0.25	0.31					
	9.9	0.19	0.17	0.16	0.17	0.16	0.24	0.18	0.19					
	11.8	0.22	0.16	0.16	0.12	0.18	0.20	0.24	0.27					

Table B5 - Maximum Penetration Rates (pg. 2 of 6)

Site No.	Exp Time yrs	Maximum Penetration Rate (mm/yr) - 2 Sample Average								Total Site Average (mm/yr)				
		Nom. 1.5 in. Pipe				Nom. 3.0 in. Pipe				Mean	Std. Dev.	Min.	Max.	Range
		a	b	e	y	B	K	M	Y					
9	1.0	1.14	0.76	0.61	0.56	0.76	0.81	0.74	0.86	0.28	0.26	0.07	1.14	1.07
	3.5	0.33	0.36	0.39	0.39	0.39	0.42	0.49	0.46					
	5.5	0.08	0.14	0.13	0.12	0.15	0.21	0.17	0.15					
	7.7	0.12	0.15	0.13	0.11	0.14	0.16	0.15	0.13					
	11.5	0.09	0.09	0.07	0.11	0.09	0.11	0.11	0.21					
	16.9	0.10	0.08	0.10	0.10	0.10	0.08	0.09	0.16					
10	1.3	0.51	0.29	0.35	0.31	0.57	0.39	0.43	0.49	0.19	0.13	0.08	0.57	0.49
	4.0	0.13	0.13	0.13	0.13	0.10	0.13	0.10	0.13					
	6.1	0.19	0.16	0.16	0.17	0.20	0.16	0.17	0.21					
	7.9	0.15	0.10	0.12	0.12	0.15	0.18	0.13	0.14					
	12.0	0.11	0.11	0.08	0.08	0.10	0.11	0.11	0.14					
11	1.4	0.89	0.69	0.58	0.51	0.69	0.60	0.69	0.76	0.32	0.19	0.12	0.89	0.77
	4.0	0.39	0.43	0.32	0.39	0.43	0.34	0.50	0.39					
	6.0	0.29	0.30	0.26	0.30	0.26	0.26	0.39	0.31					
	7.8	0.21	0.21	0.18	0.15	0.21	0.19	0.23	0.20					
	10.0	0.19	0.15	0.19	0.18	0.23	0.17	0.17	0.18					
	11.9	0.21	0.16	0.15	0.12	0.18	0.15	0.17	0.19					
12	1.9	0.13	0.13	0.13	0.13	0.13	0.13	0.13	0.13	0.16	0.08	0.05	0.40	0.35
	4.1	0.27	0.27	0.27	0.26	0.28	0.25	0.26	0.40					
	6.2	0.22	0.25	0.27	0.23	0.28	0.22	0.26	0.25					
	8.0	0.08	0.05	0.08	0.06	0.08	0.06	0.06	0.08					
	12.1	0.13	0.12	0.12	0.12	0.16	0.11	0.14	0.18					
	17.5	0.10	0.07	0.10	0.09	0.11	0.08	0.12	0.10					
13	1.9	0.48	0.43	0.64	0.48	0.63	0.87	0.84	0.95	0.42	0.22	0.16	0.95	0.79
	4.2	0.16	0.33	0.28	0.22	0.23	0.35	0.37	0.39					
	5.9	0.21	0.42	0.29	0.37	0.25	0.29	0.32	0.24					
14	1.1	0.79	0.88	0.60	0.65	0.74	0.74	0.92	1.50	0.42	0.24	0.17	1.50	1.33
	3.8	0.36	0.40	0.35	0.35	0.42	0.37	0.43	0.49					
	5.8	0.38	0.36	0.37	0.35	0.43	0.39	0.41	0.67					
	7.7	0.27	0.28	0.35	0.34	0.29	0.25	0.45	0.53					
	9.9	0.31	0.27	0.33	0.34	0.25	0.25	0.29	0.36					
	11.8	0.22	0.23	0.17	0.18	0.27	0.18	0.20	0.24					
15	2.0	0.38	0.43	0.51	0.44	0.51	0.43	0.56	0.56	0.23	0.13	0.08	0.56	0.48
	4.0	0.36	0.17	0.22	0.18	0.19	0.22	0.27	0.30					
	5.9	0.28	0.25	0.23	0.23	0.28	0.28	0.28	0.25					
	8.0	0.19	0.21	0.20	0.12	0.21	0.14	0.14	0.17					
	12.0	0.17	0.11	0.12	0.11	0.13	0.17	0.15	0.13					
	17.6	0.09	0.08	0.08	0.08	0.09	0.10	0.08	0.09					
16	2.0	0.55	0.46	0.48	0.51	0.53	0.44	0.50	0.57	0.30	0.13	0.12	0.57	0.45
	4.0	0.29	0.29	0.34	0.30	0.29	0.29	0.46	0.44					
	6.0	0.39	0.36	0.40	0.51	0.36	0.26	0.41	0.38					
	7.9	0.19	0.27	0.22	0.22	0.22	0.27	0.22	0.24					
	10.0	0.18	0.15	0.16	0.14	0.19	0.18	0.17	0.18					
	12.0	0.17	0.13	0.12	0.13	0.12	0.18	0.15	0.18					

Table B5 - Maximum Penetration Rates (pg. 3 of 6)

Site No.	Exp Time yrs	Maximum Penetration Rate (mm/yr) - 2 Sample Average								Total Site Average (mm/yr)				
		Nom. 1.5 in. Pipe				Nom. 3.0 in. Pipe				Mean	Std. Dev.	Min.	Max.	Range
		a	b	e	y	B	K	M	Y					
17	1.2	0.21	0.21	0.21	0.21	0.21	0.21	0.21	0.21	0.13	0.05	0.05	0.21	0.16
	3.8	0.15	0.11	0.15	0.13	0.15	0.16	0.21	0.14					
	5.9	0.14	0.11	0.11	0.14	0.14	0.14	0.13	0.15					
	7.7	0.11	0.12	0.12	0.10	0.13	0.13	0.12	0.13					
	11.8	0.09	0.08	0.09	0.08	0.09	0.10	0.10	0.09					
	17.0	0.06	0.06	0.06	0.05	0.06	0.07	0.07	0.09					
18	1.2	0.57	0.42	0.40	0.30	0.51	0.59	0.34	0.59	0.24	0.14	0.08	0.59	0.51
	3.8	0.27	0.27	0.32	0.23	0.23	0.35	0.29	0.25					
	5.8	0.31	0.32	0.31	0.29	0.28	0.29	0.27	0.35					
	7.7	0.18	0.12	0.13	0.14	0.18	0.19	0.14	0.16					
	9.8	0.12	0.12	0.13	0.15	0.14	0.17	0.13	0.12					
	11.7	0.11	0.09	0.09	0.08	0.09	0.15	0.10	0.10					
19	1.1	0.88	0.55	0.60	0.55	0.65	0.58	0.74	0.88	0.28	0.20	0.12	0.88	0.75
	3.7	0.26	0.31	0.34	0.26	0.25	0.32	0.41	0.34					
	5.7	0.20	0.22	0.21	0.21	0.26	0.26	0.26	0.22					
	7.6	0.15	0.13	0.15	0.13	0.19	0.19	0.19	0.22					
	9.7	0.16	0.13	0.16	0.14	0.15	0.20	0.17	0.18					
	11.6	0.12	0.14	0.16	0.14	0.14	0.19	0.13	0.14					
20	1.0	0.25	0.25	0.25	0.25	0.25	0.25	0.25	0.25	0.15	0.05	0.06	0.25	0.20
	3.6	0.14	0.14	0.14	0.14	0.14	0.14	0.14	0.14					
	5.5	0.09	0.12	0.10	0.15	0.13	0.11	0.16	0.13					
	7.7	0.09	0.13	0.13	0.14	0.11	0.12	0.14	0.21					
	9.6	0.08	0.14	0.10	0.06	0.12	0.10	0.11	0.10					
	11.6	0.15	0.11	0.16	0.14	0.10	0.18	0.12	0.14					
21	1.5	0.17	0.17	0.17	0.17	0.17	0.17	0.17	0.17	0.29	0.06	0.22	0.40	0.18
	4.0	0.39	0.25	0.32	0.40	0.38	0.25	0.30	0.28					
	6.0	0.30	0.22	0.25	0.25	0.25	0.25	0.28	0.25					
22	1.7	0.66	0.52	0.66	0.66	0.57	0.58	0.78	0.70	0.23	0.08	0.12	0.41	0.29
	3.7	0.31	0.32	0.35	0.30	0.40	0.41	0.37	0.34					
	5.6	0.31	0.25	0.26	0.24	0.28	0.28	0.28	0.27					
	7.6	0.14	0.19	0.21	0.17	0.19	0.19	0.22	0.23					
	9.6	0.17	0.16	0.14	0.15	0.18	0.17	0.17	0.18					
	11.6	0.16	0.14	0.14	0.17	0.13	0.12	0.14	0.16					
23	1.9	0.71	0.64	0.64	0.67	0.64	1.15	0.70	0.90	0.45	0.12	0.30	0.79	0.49
	4.3	0.56	0.61	0.67	0.49	0.58	0.79	0.64	0.57					
	6.2	0.59	0.51	0.37	0.41	0.37	0.45	0.41	0.53					
	8.0	0.46	0.46	0.46	0.41	0.50	0.50	0.43	0.69					
	10.2	0.36	0.36	0.36	0.36	0.34	0.38	0.34	0.41					
	12.1	0.30	0.30	0.30	0.30	0.31	0.31	0.34	0.37					
24	1.3	0.20	0.20	0.20	0.20	0.20	0.20	0.20	0.20	0.08	0.03	0.02	0.13	0.10
	4.0	0.13	0.13	0.13	0.13	0.13	0.13	0.13	0.13					
	6.1	0.08	0.08	0.08	0.08	0.09	0.08	0.08	0.12					
	7.9	0.07	0.06	0.06	0.06	0.06	0.08	0.06	0.10					
	12.0	0.06	0.05	0.04	0.05	0.06	0.06	0.06	0.07					
	17.2	0.04	0.04	0.02	0.04	0.04	0.04	0.05	0.04					

Table B5 - Maximum Penetration Rates (pg. 4 of 6)

Site No.	Exp Time yrs	Maximum Penetration Rate (mm/yr) - 2 Sample Average								Total Site Average (mm/yr)				
		Nom. 1.5 in. Pipe				Nom. 3.0 in. Pipe				Mean	Std. Dev.	Min.	Max.	Range
		a	b	e	y	B	K	M	Y					
25	1.0	0.61	0.48	0.46	0.64	0.56	0.69	0.71	0.64	0.15	0.06	0.06	0.26	0.20
	3.7	0.23	0.21	0.19	0.21	0.21	0.20	0.26	0.23					
	5.7	0.21	0.18	0.17	0.16	0.21	0.21	0.26	0.23					
	7.6	0.14	0.12	0.12	0.11	0.17	0.13	0.16	0.14					
	11.7	0.11	0.09	0.10	0.07	0.10	0.10	0.12	0.11					
	17.0	0.11	0.07	0.06	0.06	0.08	0.09	0.08	0.09					
26	1.0	0.25	0.25	0.25	0.25	0.25	0.25	0.25	0.25	0.19	0.12	0.05	0.48	0.42
	3.5	0.28	0.33	0.30	0.33	0.26	0.33	0.48	0.35					
	5.5	0.32	0.30	0.31	0.30	0.33	0.30	0.36	0.37					
	7.7	0.13	0.16	0.13	0.15	0.16	0.16	0.16	0.19					
	11.5	0.08	0.10	0.10	0.08	0.10	0.10	0.08	0.09					
	16.9	0.06	0.06	0.06	0.05	0.06	0.09	0.09	0.08					
27	2.0	0.13	0.13	0.13	0.13	0.13	0.13	0.13	0.13	0.14	0.05	0.06	0.27	0.21
	4.0	0.13	0.14	0.20	0.21	0.23	0.24	0.17	0.27					
	6.0	0.09	0.12	0.14	0.17	0.13	0.24	0.15	0.19					
	8.0	0.10	0.11	0.08	0.11	0.13	0.23	0.15	0.13					
	12.0	0.09	0.10	0.13	0.13	0.12	0.17	0.17	0.16					
	17.6	0.06	0.08	0.10	0.08	0.11	0.13	0.12	0.11					
28	1.6	0.32	0.41	0.54	0.41	0.46	0.64	0.87	0.57	0.38	0.09	0.23	0.57	0.34
	5.6	0.23	0.31	0.27	0.25	0.25	0.34	0.25	0.34					
	7.7	0.45	0.41	0.39	0.35	0.44	0.53	0.39	0.50					
	9.6	0.38	0.35	0.36	0.38	0.44	0.48	0.40	0.57					
29	2.0	0.61	0.47	0.69	0.71	0.53	0.86	0.55	0.47	0.34	0.09	0.19	0.57	0.38
	4.1	0.40	0.29	0.37	0.42	0.45	0.48	0.37	0.56					
	6.0	0.44	0.33	0.39	0.35	0.24	0.34	0.38	0.34					
	8.0	0.33	0.22	0.29	0.29	0.31	0.29	0.41	0.57					
	10.0	0.37	0.25	0.19	0.23	0.21	0.30	0.25	0.45					
	12.0	0.31	0.19	0.29	0.31	0.28	0.46	0.25	0.46					
30	1.1	0.55	0.23	0.28	0.23	0.28	0.37	0.37	0.42	0.12	0.03	0.06	0.19	0.13
	3.6	0.14	0.14	0.14	0.14	0.14	0.14	0.14	0.14					
	5.7	0.09	0.09	0.11	0.07	0.12	0.12	0.12	0.12					
	8.2	0.14	0.11	0.10	0.11	0.19	0.11	0.15	0.10					
	11.6	0.12	0.11	0.13	0.11	0.12	0.12	0.14	0.14					
	17.0	0.07	0.07	0.08	0.06	0.08	0.10	0.11	0.10					
31	2.0	0.38	0.42	0.85	0.50	0.47	0.61	0.66	0.69	0.13	0.06	0.06	0.29	0.23
	4.1	0.15	0.16	0.15	0.15	0.19	0.29	0.17	0.21					
	6.0	0.12	0.11	0.13	0.22	0.12	0.24	0.14	0.19					
	8.0	0.11	0.14	0.12	0.12	0.13	0.14	0.20	0.11					
	12.0	0.07	0.06	0.07	0.06	0.09	0.09	0.08	0.08					
	17.7	0.07	0.06	0.06	0.07	0.06	0.13	0.09	0.07					
32	1.0	1.02	0.66	0.53	0.56	0.76	0.58	0.69	0.66	0.17	0.05	0.09	0.30	0.21
	3.7	0.30	0.27	0.21	0.19	0.27	0.23	0.25	0.28					
	5.8	0.16	0.14	0.13	0.16	0.17	0.17	0.16	0.18					
	7.6	0.11	0.14	0.14	0.13	0.13	0.16	0.15	0.15					
	9.6	0.11	0.15	0.12	0.12	0.14	0.13	0.14	0.25					
	11.7	0.13	0.09	0.10	0.11	0.13	0.19	0.13	0.20					

Table B5 - Maximum Penetration Rates (pg. 5 of 6)

Site No.	Exp Time yrs	Maximum Penetration Rate (mm/yr) - 2 Sample Average								Total Site Average (mm/yr)				
		Nom. 1.5 in. Pipe				Nom. 3.0 in. Pipe				Mean	Std. Dev.	Min.	Max.	Range
		a	b	e	y	B	K	M	Y					
33	1.0	0.25	0.25	0.25	0.25	0.25	0.25	0.25	0.25	0.23	0.07	0.09	0.43	0.35
	3.7	0.26	0.14	0.14	0.17	0.22	0.21	0.27	0.26					
	5.8	0.20	0.16	0.09	0.22	0.16	0.24	0.14	0.12					
	7.6	0.43	0.20	0.22	0.32	0.33	0.39	0.32	0.31					
	9.7	0.30	0.21	0.24	0.19	0.29	0.27	0.30	0.29					
	11.7	0.24	0.21	0.20	0.23	0.23	0.23	0.21	0.18					
34	1.4	0.36	0.38	0.36	0.33	0.40	0.49	0.40	0.53	0.15	0.05	0.07	0.27	0.20
	4.0	0.15	0.13	0.13	0.11	0.10	0.11	0.10	0.11					
	6.1	0.20	0.12	0.07	0.13	0.15	0.16	0.12	0.20					
	8.0	0.13	0.15	0.14	0.17	0.14	0.21	0.13	0.22					
	9.9	0.21	0.12	0.22	0.24	0.18	0.19	0.20	0.27					
	12.0	0.12	0.08	0.09	0.07	0.12	0.08	0.11	0.16					
35	1.9	0.13	0.23	0.13	0.13	0.13	0.13	0.23	0.24	0.09	0.05	0.01	0.31	0.29
	4.1	0.10	0.12	0.12	0.12	0.12	0.11	0.14	0.15					
	6.2	0.08	0.08	0.16	0.08	0.09	0.08	0.07	0.09					
	8.0	0.10	0.08	0.07	0.06	0.06	0.12	0.09	0.31					
	12.1	0.02	0.05	0.04	0.04	0.08	0.05	0.14	0.06					
	17.5	0.02	0.08	0.01	0.02	0.05	0.05	0.05	0.09					
36	2.0	0.36	0.47	0.43	0.48	0.47	0.61	0.55	0.47	0.15	0.06	0.06	0.31	0.25
	4.1	0.23	0.24	0.19	0.19	0.24	0.31	0.28	0.26					
	6.0	0.20	0.20	0.16	0.17	0.17	0.22	0.20	0.18					
	8.0	0.15	0.17	0.11	0.12	0.15	0.18	0.16	0.16					
	12.0	0.12	0.10	0.10	0.10	0.10	0.13	0.10	0.09					
	17.7	0.07	0.07	0.08	0.06	0.07	0.09	0.07	0.08					
37	2.0	0.48	0.44	0.61	0.61	0.53	0.55	0.67	1.14	0.23	0.09	0.13	0.57	0.44
	4.1	0.28	0.26	0.38	0.41	0.37	0.36	0.34	0.57					
	6.0	0.19	0.22	0.16	0.31	0.16	0.20	0.24	0.25					
	8.0	0.18	0.16	0.17	0.17	0.20	0.22	0.21	0.30					
	10.1	0.17	0.18	0.22	0.19	0.19	0.18	0.21	0.29					
	12.0	0.16	0.14	0.17	0.14	0.17	0.13	0.20	0.27					
38	1.4	0.18	0.18	0.18	0.18	0.18	0.18	0.18	0.18	0.09	0.04	0.04	0.18	0.14
	4.0	0.13	0.18	0.13	0.13	0.13	0.13	0.13	0.13					
	6.1	0.17	0.11	0.08	0.08	0.15	0.07	0.10	0.15					
	8.0	0.17	0.09	0.06	0.08	0.11	0.08	0.09	0.08					
	12.0	0.07	0.07	0.06	0.04	0.06	0.05	0.09	0.06					
	17.2	0.06	0.05	0.04	0.05	0.05	0.06	0.05	0.05					
39	1.4	0.18	0.18	0.18	0.18	0.18	0.18	0.18	0.18	0.17	0.05	0.10	0.29	0.19
	4.0	0.27	0.15	0.24	0.16	0.10	0.15	0.20	0.29					
	6.1	0.13	0.18	0.16	0.21	0.20	0.16	0.18	0.21					
	8.0	0.15	0.16	0.15	0.13	0.21	0.23	0.17	0.24					
	9.9	0.17	0.11	0.11	0.13	0.14	0.17	0.13	0.23					
	12.0	0.16	0.12	0.11	0.13	0.15	0.15	0.20	0.22					
40	2.0	0.58	0.36	0.48	0.39	0.36	0.52	0.41	0.53	0.24	0.07	0.14	0.38	0.23
	4.1	0.38	0.32	0.27	0.22	0.25	0.28	0.27	0.28					
	6.0	0.26	0.29	0.30	0.25	0.28	0.37	0.31	0.36					
	8.0	0.27	0.32	0.20	0.23	0.22	0.31	0.30	0.29					
	10.0	0.19	0.16	0.22	0.16	0.17	0.18	0.16	0.22					
	12.0	0.29	0.14	0.15	0.17	0.15	0.19	0.17	0.16					

Table B5 - Maximum Penetration Rates (pg. 6 of 6)

Site No.	Exp Time yrs	Maximum Penetration Rate (mm/yr) - 2 Sample Average								Total Site Average (mm/yr)				
		Nom. 1.5 in. Pipe				Nom. 3.0 in. Pipe				Mean	Std. Dev.	Min.	Max.	Range
		a	b	e	y	B	K	M	Y					
41	1.5	0.32	0.66	0.56	0.68	0.75	0.51	0.71	0.54	0.20	0.05	0.10	0.30	0.20
	4.0	0.22	0.23	0.25	0.29	0.30	0.24	0.27	0.29					
	6.0	0.22	0.21	0.20	0.25	0.23	0.22	0.28	0.22					
	7.9	0.20	0.21	0.18	0.19	0.19	0.19	0.20	0.18					
	12.0	0.21	0.20	0.17	0.19	0.18	0.15	0.18	0.17					
	17.4	0.18	0.14	0.13	0.15	0.13	0.10	0.15	0.11					
42	2.0	0.67	0.74	0.75	0.64	0.69	0.79	0.99	0.70	0.32	0.10	0.18	0.53	0.34
	4.1	0.40	0.46	0.45	0.47	0.49	0.53	0.51	0.46					
	6.0	0.33	0.39	0.36	0.31	0.33	0.36	0.40	0.34					
	8.0	0.24	0.28	0.36	0.35	0.25	0.30	0.28	0.29					
	10.1	0.21	0.21	0.26	0.23	0.24	0.32	0.26	0.26					
	12.0	0.20	0.19	0.23	0.18	0.19	0.26	0.19	0.25					
43	1.3	0.35	0.70	0.47	0.55	0.86	0.88	1.29	0.92	0.30	0.12	0.15	0.63	0.47
	4.1	0.24	0.27	0.22	0.28	0.63	0.49	0.51	0.37					
	6.2	0.36	0.34	0.27	0.31	0.43	0.54	0.49	0.22					
	8.0	0.27	0.32	0.24	0.22	0.33	0.43	0.25	0.29					
	9.9	0.24	0.19	0.18	0.19	0.35	0.26	0.30	0.40					
	12.0	0.19	0.17	0.21	0.22	0.29	0.17	0.16	0.15					
44	1.1	0.88	0.83	0.65	0.74	0.60	0.90	0.74	0.88	0.25	0.10	0.12	0.55	0.43
	3.6	0.55	0.30	0.38	0.39	0.32	0.31	0.31	0.35					
	5.7	0.31	0.23	0.23	0.29	0.25	0.32	0.27	0.30					
	7.6	0.24	0.16	0.21	0.17	0.19	0.17	0.21	0.29					
	11.6	0.19	0.12	0.14	0.15	0.14	0.13	0.18	0.16					
45	1.2	0.21	0.42	0.32	0.28	0.21	0.36	0.64	0.59	0.23	0.07	0.15	0.41	0.26
	3.8	0.24	0.19	0.16	0.16	0.21	0.24	0.21	0.33					
	5.8	0.20	0.19	0.18	0.15	0.21	0.16	0.21	0.23					
	7.7	0.16	0.15	0.20	0.15	0.18	0.20	0.20	0.22					
	9.8	0.37	0.30	0.36	0.30	0.31	0.36	0.33	0.41					
	11.7	0.18	0.17	0.18	0.18	0.18	0.24	0.21	0.27					
46	1.5	0.97	0.91	0.93	0.91	0.85	0.68	0.95	1.34	0.31	0.15	0.10	0.70	0.60
	4.0	0.51	0.41	0.50	0.52	0.42	0.37	0.67	0.70					
	5.1	0.34	0.32	0.33	0.27	0.34	0.23	0.48	0.48					
	8.0	0.19	0.25	0.34	0.37	0.22	0.22	0.43	0.43					
	10.2	0.18	0.24	0.17	0.21	0.20	0.16	0.21	0.20					
	12.0	0.10	0.13	0.14	0.22	0.16	0.13	0.24	0.17					
47	1.5	0.17	0.17	0.17	0.17	0.17	0.17	0.17	0.17	0.08	0.03	0.03	0.12	0.09
	4.1	0.12	0.12	0.12	0.12	0.12	0.12	0.12	0.12					
	6.1	0.08	0.08	0.08	0.08	0.08	0.08	0.08	0.08					
	8.0	0.06	0.06	0.06	0.06	0.06	0.06	0.06	0.06					
	12.1	0.03	0.10	0.05	0.07	0.07	0.05	0.06	0.05					
	17.4	0.06	0.08	0.05	0.08	0.07	0.06	0.07	0.07					

Table B6 - Ratio of Maximum Observed Penetration to Average Penetration from Mass Loss (pg. 1 of 6)

Site No.	Exp Time yrs	2 Sample Average Pitting Ratio (m/m)								Averge Site Pitting Ratio (m/m)				
		Nom. 1.5 in. Pipe				Nom. 3.0 in. Pipe				Mean	Std. Dev.	Min.	Max.	Range
		a	b	e	y	B	K	M	Y					
1	1.0	5.9	5.5	5.9	5.9	5.5	4.7	5.0	5.5	6.2	1.1	4.5	9.3	4.8
	3.6	9.1	5.4	6.7	6.7	6.5	5.6	6.8	8.6					
	5.5	8.2	5.4	4.5	7.7	7.0	6.7	6.4	7.0					
	7.7	5.8	7.1	5.0	5.1	5.3	6.3	5.5	4.5					
	9.6	5.2	5.1	5.8	6.8	6.7	6.1	6.6	7.0					
	11.6	6.8	5.4	7.9	4.8	5.2	6.5	5.7	9.3					
2	2.1	3.4	3.3	3.6	2.7	3.4	3.0	3.0	3.6	6.8	2.6	2.7	13.7	10.9
	4.0	12.4	8.9	8.3	10.2	9.2	10.9	9.3	12.3					
	5.9	7.2	6.2	7.7	7.9	6.7	7.0	6.5	9.3					
	7.9	8.3	6.4	7.2	9.1	8.1	9.5	8.4	13.7					
	12.0	6.1	4.4	6.2	8.9	5.8	6.5	4.9	7.4					
	17.6	6.0	4.7	4.7	5.2	3.8	5.9	5.3	5.9					
3	2.0	27.0	16.7	16.9	18.3	22.5	17.3	15.6	23.1	13.4	3.9	7.1	27.0	20.0
	4.1	16.7	13.5	15.5	12.0	11.4	10.5	13.3	15.1					
	6.0	14.1	9.3	12.4	13.5	12.4	14.9	12.7	10.1					
	8.0	11.0	8.6	10.9	12.4	11.0	10.9	9.6	11.7					
	10.1	18.4	10.0	11.0	12.0	15.1	11.3	14.5	13.1					
	12.1	15.7	10.7	8.6	7.1	10.8	10.6	11.0	12.2					
4	1.4	4.4	5.5	4.4	4.7	4.7	4.7	4.7	4.4	6.7	2.5	4.1	13.6	9.4
	4.0	5.8	5.0	4.9	4.5	4.6	5.7	5.6	7.4					
	6.1	11.9	7.0	6.5	6.7	7.0	8.0	6.8	12.0					
	8.0	7.5	4.4	4.1	5.4	4.5	6.9	5.0	8.6					
	12.0	13.6	7.7	7.4	8.7	8.0	10.1	8.0	13.3					
5	1.9	5.9	7.3	5.0	4.7	6.5	6.5	7.3	6.5	6.0	2.1	3.1	12.6	9.5
	4.1	12.6	6.5	9.4	8.7	8.3	9.7	8.9	10.0					
	6.2	5.9	4.6	7.3	4.6	5.3	10.4	8.5	5.7					
	8.1	4.0	5.1	5.8	4.8	4.8	6.0	6.5	5.5					
	12.1	3.3	4.3	3.9	4.5	4.5	4.3	4.1	4.5					
	17.5	6.9	3.1	4.7	3.3	4.6	5.4	4.7	5.0					
6	1.9	45.8	85.1	170.1	65.4	143.9	58.9	62.2	68.7	27.0	32.7	5.9	170.1	164.2
	4.1	20.6	20.6	15.3	16.4	20.9	20.6	30.1	20.9					
	6.2	24.0	15.3	13.1	19.6	25.1	20.4	16.0	14.7					
	8.1	13.9	14.5	7.9	13.9	14.5	18.0	13.1	13.8					
	12.1	15.0	10.9	8.4	13.1	17.8	9.5	11.3	16.1					
	17.5	8.0	12.3	5.9	8.7	11.0	10.3	6.9	8.5					
7	1.0	13.1	9.3	10.9	10.9	10.9	13.1	9.3	10.9	6.7	2.6	3.4	13.1	9.7
	3.5	7.9	5.2	6.0	4.4	4.4	7.2	5.2	5.0					
	7.7	3.8	5.2	3.7	4.4	5.6	7.4	6.1	3.4					
	11.5	6.8	5.1	6.7	4.4	6.4	5.9	5.6	5.6					
	16.9	4.1	4.3	5.6	6.2	8.6	7.7	6.2	7.2					
8	1.1	41.1	28.0	35.5	24.5	35.5	46.9	38.4	57.9	16.0	11.4	5.8	57.9	52.1
	3.8	15.8	12.4	14.4	12.1	14.6	19.6	16.5	20.7					
	5.8	12.7	10.6	11.7	13.6	15.3	16.3	11.1	13.9					
	7.7	12.3	10.3	13.5	10.6	9.9	16.1	11.7	14.3					
	9.9	9.5	8.5	8.7	9.5	7.9	11.5	8.4	8.8					
	11.8	7.8	7.2	6.3	5.8	6.2	7.6	8.7	9.4					

Table B6 - Ratio of Maximum Observed Penetration to Average Penetration from Mass Loss (pg. 2 of 6)

Site No.	Exp Time yrs	2 Sample Average Pitting Ratio (m/m)							Averge Site Pitting Ratio (m/m)				
		Nom. 1.5 in. Pipe				Nom. 3.0 in. Pipe				Mean	Std. Dev.	Min.	Max.	Range
		abey	B	K	M	Y							
9	1.0	49.1	16.4	22.4	16.0	32.7	34.9	31.6	44.5	13.8	10.6	4.0	49.1	45.0
	3.5	17.7	14.2	15.4	17.7	18.6	31.6	23.4	24.6					
	5.5	5.4	8.8	7.6	7.7	8.1	10.9	8.7	8.1					
	7.7	7.4	8.9	7.8	7.0	8.0	9.2	9.4	8.4					
	11.5	5.8	5.7	4.0	6.8	5.3	6.3	7.0	13.4					
	16.9	8.4	5.6	7.2	7.9	8.1	6.8	8.0	13.7					
10	1.3	12.2	8.2	9.8	8.1	17.2	13.1	10.3	13.6	8.0	2.7	3.8	17.2	13.5
	4.0	9.3	6.2	4.8	10.9	6.2	10.9	3.8	8.2					
	6.1	8.9	7.3	7.3	8.1	9.2	6.7	7.9	9.6					
	7.9	7.1	4.9	5.5	5.4	6.7	8.0	5.7	6.0					
	12.0	7.6	6.3	5.3	5.7	6.1	7.7	7.2	8.5					
11	1.4	106.9	49.7	52.3	36.6	62.2	54.0	62.2	68.7	35.9	16.0	18.2	106.9	88.7
	4.0	29.0	39.8	29.7	33.3	44.5	39.3	51.0	45.1					
	6.0	40.4	33.2	28.5	35.2	31.2	28.5	46.3	31.8					
	7.8	41.9	19.0	19.3	20.1	28.8	27.1	29.4	29.0					
	10.0	35.0	23.1	24.9	22.9	34.6	24.0	27.0	31.0					
	11.9	38.1	18.2	22.1	20.0	26.8	22.9	30.8	26.2					
12	1.9	32.7	32.7	16.4	21.8	32.7	32.7	21.8	32.7	15.7	7.5	5.6	32.7	27.1
	4.1	20.6	14.8	18.0	13.7	20.1	17.4	21.1	26.6					
	6.2	11.0	17.1	12.5	12.2	13.5	12.6	15.3	13.5					
	8.0	17.0	7.9	18.9	13.1	17.4	14.7	14.7	14.9					
	12.1	11.5	8.6	8.8	9.9	12.1	9.5	11.8	16.9					
	17.5	6.4	5.6	7.9	7.5	9.2	7.6	10.6	7.7					
13	1.9	9.4	7.0	8.5	6.9	7.9	9.5	12.5	11.3	8.1	2.6	4.2	16.5	12.3
	4.2	6.5	5.1	7.9	7.3	8.3	16.5	9.2	8.9					
	5.9	4.9	7.6	6.5	7.5	7.0	7.2	6.3	4.2					
14	1.1	74.1	62.2	42.5	36.6	52.3	41.9	65.4	106.3	26.4	18.9	10.1	106.3	96.2
	3.8	17.7	15.7	13.1	14.4	17.9	16.7	19.0	24.2					
	5.8	23.7	14.5	15.3	16.7	18.9	18.4	19.8	34.5					
	7.7	26.8	15.2	28.0	30.6	30.6	29.6	26.8	45.8					
	9.9	20.7	10.8	18.9	16.2	12.7	15.9	16.2	25.8					
	11.8	16.1	15.8	10.1	12.2	17.3	12.6	13.4	15.6					
15	2.0	8.2	10.6	11.4	10.9	11.9	9.7	10.7	10.7	6.7	2.5	2.7	11.9	9.2
	4.0	10.2	5.7	7.6	5.7	6.3	6.5	7.4	10.1					
	5.9	6.9	6.3	6.3	6.4	6.8	7.3	8.1	5.9					
	8.0	7.7	9.2	7.5	4.5	7.3	5.2	5.6	6.8					
	12.0	6.3	3.8	4.8	4.5	5.0	6.5	5.8	4.5					
	17.6	3.4	3.0	3.4	3.6	3.2	2.7	3.0	5.3					
16	2.0	14.8	9.4	9.2	12.5	12.5	10.0	9.8	14.0	9.5	3.0	4.7	18.3	13.5
	4.0	10.4	7.7	10.1	10.2	10.4	10.0	14.3	13.1					
	6.0	11.6	11.2	14.0	18.3	9.8	9.0	12.6	11.8					
	7.9	7.6	11.0	9.7	9.7	9.8	11.9	10.0	9.5					
	10.0	6.7	6.0	6.2	5.9	8.3	7.4	6.3	7.5					
	12.0	6.5	4.9	4.7	5.6	4.8	7.4	6.1	7.2					

Table B6 - Ratio of Maximum Observed Penetration to Average Penetration from Mass Loss (pg. 3 of 6)

Site No.	Exp Time yrs	2 Sample Average Pitting Ratio (m/m)								Average Site Pitting Ratio (m/m)				
		Nom. 1.5 in. Pipe				Nom. 3.0 in. Pipe				Mean	Std. Dev.	Min.	Max.	Range
		abey	BKMY											
17	1.2	4.4	4.4	4.7	4.7	4.1	4.4	4.1	4.4	3.6	0.7	2.6	5.8	3.2
	3.8	4.4	3.3	4.2	4.1	4.1	4.6	5.8	3.8					
	5.9	4.0	2.9	3.0	4.0	3.4	3.4	3.4	3.7					
	7.7	3.3	3.3	3.7	3.0	3.5	3.6	3.2	3.4					
	11.8	2.9	2.9	3.3	2.8	2.6	3.0	3.3	2.7					
	17.0	3.5	2.8	2.8	2.7	3.0	3.5	3.3	4.2					
18	1.2	44.2	18.7	31.1	15.3	39.3	61.1	20.9	36.6	17.4	10.2	7.3	61.1	53.7
	3.8	21.8	14.5	16.5	13.9	17.1	28.4	23.4	14.2					
	5.8	20.2	11.8	14.5	16.9	14.4	16.0	12.7	20.1					
	7.7	20.0	7.3	9.0	13.7	16.7	19.6	11.9	15.6					
	9.8	10.7	8.6	10.4	11.3	11.8	12.3	10.6	10.4					
	11.7	11.3	10.2	8.7	9.6	13.4	17.0	12.0	7.4					
19	1.1	41.4	19.6	24.3	22.4	30.5	23.4	26.2	31.1	16.0	6.3	8.7	41.4	32.7
	3.7	12.4	19.6	14.9	11.3	10.2	13.1	21.4	16.4					
	5.7	13.7	10.9	14.3	13.7	16.5	16.8	19.0	15.6					
	7.6	11.5	8.7	11.1	9.3	15.3	15.8	12.2	17.3					
	9.7	13.5	10.4	12.5	12.4	11.8	17.6	13.7	14.8					
	11.6	12.6	11.6	12.2	12.7	13.5	16.4	11.9	11.8					
20	1.0	7.3	8.2	5.9	6.5	7.3	6.5	7.3	7.3	6.9	1.8	3.5	15.2	11.7
	3.6	5.9	5.5	7.3	5.7	5.9	5.9	6.2	6.5					
	5.5	5.5	6.3	6.0	10.0	8.2	6.8	7.9	8.3					
	7.7	6.8	7.3	9.7	11.0	6.5	7.9	7.9	15.2					
	9.6	5.4	7.1	5.4	3.5	5.6	5.1	5.7	5.6					
	11.6	6.6	4.8	7.7	7.0	5.3	8.6	6.1	7.2					
21	1.5	3.3	3.1	3.1	3.1	3.0	3.3	3.3	3.1	10.5	3.0	7.6	17.2	9.6
	4.0	16.0	10.1	11.3	17.2	14.0	8.8	10.8	11.5					
	6.0	11.1	7.7	7.9	8.1	8.2	7.6	9.4	8.5					
22	1.7	20.6	14.3	14.4	15.2	17.8	17.0	24.3	17.1	7.1	1.8	4.7	10.9	6.2
	3.7	9.2	7.7	9.8	7.8	10.5	10.6	10.1	9.3					
	5.6	8.4	6.7	8.2	6.5	8.8	8.8	9.0	10.9					
	7.6	4.9	5.9	5.6	5.9	5.7	6.1	5.5	6.0					
	9.6	6.3	5.8	4.7	5.8	6.2	6.3	5.5	6.8					
	11.6	6.7	5.7	5.8	7.2	5.0	5.3	5.5	6.6					
23	1.9	6.4	4.1	4.1	4.2	5.4	8.7	5.1	5.6	5.1	0.7	4.1	7.4	3.3
	4.3	5.5	5.1	5.4	5.2	5.3	7.4	5.9	6.5					
	6.2	6.9	5.2	4.8	4.2	4.6	4.9	4.7	4.9					
	8.0	5.2	4.6	6.1	4.5	5.6	4.9	4.6	5.9					
	10.2	4.9	4.4	4.9	5.0	4.5	5.0	4.1	5.3					
	12.1	4.7	4.5	4.8	4.7	4.7	5.0	5.4	4.8					
24	1.3	32.7	21.8	21.8	21.8	32.7	21.8	21.8	32.7	17.3	8.5	5.8	43.6	37.8
	4.0	32.7	21.8	26.2	32.7	32.7	32.7	32.7	43.6					
	6.1	14.5	10.9	10.1	16.4	12.5	11.9	11.9	18.3					
	7.9	16.0	13.1	13.1	16.4	12.4	18.2	13.1	25.4					
	12.0	14.1	10.7	9.2	11.7	15.3	13.1	14.0	16.9					
	17.2	13.1	8.3	5.8	12.2	14.0	13.6	14.7	15.3					

Table B6 - Ratio of Maximum Observed Penetration to Average Penetration from Mass Loss (pg. 4 of 6)

Site No.	Exp Time yrs	2 Sample Average Pitting Ratio (m/m)								Averge Site Pitting Ratio (m/m)				
		Nom. 1.5 in. Pipe				Nom. 3.0 in. Pipe				Mean	Std. Dev.	Min.	Max.	Range
		a	b	e	y	B	K	M	Y					
25	1.0	39.3	20.7	29.4	40.9	48.0	44.2	45.8	40.9	14.8	6.0	7.0	37.1	30.0
	3.7	37.1	14.0	16.1	17.8	29.0	21.1	17.8	15.9					
	5.7	18.5	15.8	13.4	12.4	17.1	18.5	22.3	22.7					
	7.6	15.3	10.0	11.8	11.1	17.9	13.8	17.1	16.2					
	11.7	10.0	7.0	9.2	7.2	11.2	8.4	11.3	12.2					
	17.0	16.4	7.8	7.2	8.9	12.6	12.7	13.6	13.8					
26	1.0	8.2	7.3	8.2	8.2	8.2	8.2	7.3	7.3	16.1	9.8	5.4	39.3	33.9
	3.5	15.9	13.4	15.3	16.7	12.4	15.5	19.6	16.5					
	5.5	30.5	20.6	25.8	27.9	39.3	36.0	36.5	37.4					
	7.7	19.6	14.6	12.4	18.0	20.5	20.9	19.6	22.3					
	11.5	7.1	7.3	8.0	6.5	7.2	8.1	6.5	8.3					
	16.9	7.0	5.4	6.4	5.7	5.8	9.6	7.9	8.1					
27	2.0	16.4	10.9	13.1	13.1	10.9	10.9	10.9	13.1	5.9	2.0	2.7	11.8	9.1
	4.0	4.1	4.6	5.2	6.0	6.7	7.1	5.4	10.2					
	6.0	4.6	4.8	5.1	8.7	6.3	11.8	7.6	9.4					
	8.0	4.4	5.2	4.1	4.4	6.2	9.6	7.7	6.2					
	12.0	3.9	3.7	4.5	5.2	4.3	7.5	6.5	6.0					
	17.6	2.7	3.3	4.5	3.8	5.0	7.1	6.2	6.1					
28	1.6	3.8	6.3	6.2	5.7	9.0	8.2	12.4	6.7	5.4	1.3	3.2	8.0	4.8
	5.6	3.2	4.9	4.1	3.4	4.1	4.6	3.5	4.6					
	7.7	7.7	6.3	5.3	4.6	5.6	6.4	5.8	5.5					
	9.6	6.0	5.6	5.5	5.6	6.5	7.4	6.4	8.0					
29	2.0	7.9	7.3	9.3	8.3	7.9	10.6	6.9	5.8	5.5	1.3	3.4	7.9	4.5
	4.1	5.9	4.0	5.4	6.5	6.5	7.0	5.7	7.5					
	6.0	6.8	5.7	6.4	5.5	4.0	5.5	6.6	5.7					
	8.0	4.8	3.4	4.3	4.2	5.9	3.6	7.6	7.4					
	10.0	5.9	4.7	3.5	4.3	3.7	5.1	4.4	7.5					
	12.0	4.4	3.6	4.6	4.9	5.6	7.9	5.3	7.0					
30	1.1	17.4	7.3	7.1	7.3	8.7	9.5	9.5	11.8	7.2	1.9	4.9	11.9	7.0
	3.6	11.9	10.9	10.9	10.9	9.3	10.1	9.3	10.9					
	5.7	7.3	5.7	6.0	5.2	7.3	6.8	7.1	7.6					
	8.2	7.3	5.0	5.3	5.3	8.6	5.7	7.1	5.3					
	11.6	6.8	6.0	7.9	6.3	5.5	6.1	7.4	8.3					
	17.0	5.4	4.9	6.0	5.1	5.8	7.0	7.2	6.6					
31	2.0	28.0	16.6	36.5	17.0	30.3	26.2	26.2	32.1	10.1	3.3	4.3	19.3	15.0
	4.1	7.9	10.0	7.8	7.4	9.3	12.3	8.7	9.7					
	6.0	11.4	9.5	10.3	18.2	12.6	19.3	9.1	13.1					
	8.0	13.1	13.1	10.1	8.9	13.1	13.4	18.2	9.1					
	12.0	7.7	6.3	7.9	7.3	9.6	9.6	9.3	6.9					
	17.7	6.7	6.5	7.6	7.7	4.3	13.1	10.5	7.0					
32	1.0	130.8	34.0	34.3	36.0	65.4	37.6	44.2	56.7	12.8	3.9	7.9	24.0	16.1
	3.7	24.0	14.5	12.3	12.6	18.7	17.1	18.1	19.2					
	5.8	13.4	10.0	8.5	13.1	10.8	11.6	9.3	11.0					
	7.6	10.5	9.8	10.2	11.1	9.9	12.1	12.5	11.1					
	9.6	13.4	11.2	9.7	11.5	11.3	13.4	12.6	23.7					
	11.7	12.2	7.9	9.1	8.8	10.7	15.6	9.7	19.0					

Table B6 - Ratio of Maximum Observed Penetration to Average Penetration from Mass Loss (pg. 5 of 6)

Site No.	Exp Time yrs	2 Sample Average Pitting Ratio (m/m) Nom. 1.5 in. Pipe abeyBKMY				Nom. 3.0 in. Pipe				Averge Site Pitting Ratio (m/m) Mean	Std. Dev.	Min.	Max.	Range
33	1.0	16.4	16.4	16.4	13.1	16.4	16.4	16.4	16.4	5.8	1.5	2.8	10.1	7.3
	3.7	7.1	4.0	3.8	4.7	5.7	5.6	7.3	6.5					
	5.8	6.4	5.1	2.8	6.0	5.5	6.3	4.5	6.1					
	7.6	10.1	4.3	4.7	7.2	7.7	8.7	7.1	7.9					
	9.7	6.7	4.3	5.3	4.9	7.3	6.7	6.6	6.9					
	11.7	5.2	4.6	4.1	4.8	4.8	6.1	4.9	3.9					
34	1.4	10.1	10.6	10.9	10.7	10.3	12.6	10.3	15.8	8.7	2.9	3.7	15.5	11.7
	4.0	9.4	7.6	7.3	6.2	5.8	6.2	5.8	6.2					
	6.1	10.1	6.5	3.7	7.8	8.4	8.0	6.8	10.8					
	8.0	9.2	9.1	8.7	9.7	8.0	13.7	7.9	13.9					
	9.9	12.8	6.5	11.9	15.0	10.3	12.2	11.7	15.5					
	12.0	10.0	4.9	6.7	6.5	6.1	5.5	6.2	10.1					
35	1.9	8.2	12.4	9.3	9.3	8.2	7.3	10.1	13.1	10.2	5.4	1.6	27.6	26.0
	4.1	6.2	6.5	7.3	9.3	6.2	5.6	7.6	8.7					
	6.2	10.9	10.9	17.4	13.1	1.6	18.7	11.8	9.8					
	8.0	8.7	7.5	8.9	7.7	5.6	11.8	8.7	27.6					
	12.1	4.1	5.8	5.4	11.9	13.1	8.5	18.1	9.3					
	17.5	5.2	8.6	3.9	12.4	17.4	24.7	7.4	12.7					
36	2.0	15.3	20.2	15.9	22.6	30.3	39.3	28.1	26.9	14.1	4.1	8.0	23.7	15.7
	4.1	17.3	12.4	8.9	9.7	15.0	19.2	18.8	18.3					
	6.0	23.7	18.5	14.2	14.9	19.2	18.9	22.4	17.2					
	8.0	15.7	11.8	9.2	10.4	17.4	19.0	15.6	15.6					
	12.0	14.7	9.1	9.8	10.8	13.4	14.0	12.3	11.5					
	17.7	10.1	8.0	8.8	8.0	13.1	13.1	12.3	12.0					
37	2.0	10.8	10.0	13.1	13.1	11.4	11.7	13.9	24.5	7.3	1.6	4.7	12.0	7.4
	4.1	7.2	6.2	9.2	10.3	8.6	8.6	8.0	12.0					
	6.0	7.2	7.3	6.9	10.4	6.5	7.5	8.9	10.2					
	8.0	6.5	5.3	6.5	6.8	6.6	7.3	7.6	9.7					
	10.1	5.0	5.4	7.3	6.3	6.0	5.4	6.5	7.5					
	12.0	6.0	4.7	6.6	6.5	6.4	4.9	6.9	8.0					
38	1.4	32.7	32.7	32.7	32.7	32.7	21.8	21.8	21.8	13.7	7.7	5.0	32.7	27.7
	4.0	26.2	26.2	18.7	21.8	26.2	32.7	21.8	26.2					
	6.1	29.1	21.3	10.9	16.4	22.0	8.6	13.1	19.1					
	8.0	17.0	9.2	6.2	8.7	11.1	8.5	8.6	7.5					
	12.0	10.9	9.3	6.1	5.0	7.6	5.8	12.5	6.3					
	17.2	9.7	6.3	8.0	9.1	9.7	8.9	7.9	8.6					
39	1.4	6.5	5.5	5.5	5.0	5.5	5.0	5.5	3.1	8.2	2.1	3.3	13.9	10.6
	4.0	8.9	5.6	8.6	6.1	3.3	5.2	6.5	10.4					
	6.1	8.8	9.0	9.6	12.1	9.8	8.0	9.0	10.2					
	8.0	7.7	7.3	6.5	7.6	8.8	9.8	6.9	9.9					
	9.9	9.7	5.3	5.6	7.4	6.5	8.9	7.2	12.0					
	12.0	10.5	6.7	5.4	7.5	7.2	8.1	9.6	13.9					
40	2.0	14.3	8.7	11.8	10.1	8.0	10.7	9.5	9.8	7.5	1.6	4.5	13.2	8.7
	4.1	10.0	7.9	7.2	6.2	6.4	7.0	6.3	7.2					
	6.0	7.4	8.2	9.2	7.4	7.0	8.7	7.9	9.8					
	8.0	8.8	10.2	6.4	8.8	6.5	9.1	7.7	7.8					
	10.0	7.1	5.6	7.8	6.8	5.3	6.5	4.5	7.5					
	12.0	13.2	5.6	6.3	8.0	5.2	6.9	7.3	5.9					

Table B6 - Ratio of Maximum Observed Penetration to Average Penetration from Mass Loss (pg. 6 of 6)

Site No.	Exp Time yrs	2 Sample Average Pitting Ratio (m/m)								Averge Site Pitting Ratio (m/m)				
		Nom. 1.5 in. Pipe				Nom. 3.0 in. Pipe				Mean	Std. Dev.	Min.	Max.	Range
		a	b	e	y	B	K	M	Y					
41	1.5	15.5	51.0	43.2	26.2	48.0	32.7	68.7	52.3	9.1	1.6	6.5	13.5	7.0
	4.0	8.2	6.5	8.7	9.8	11.4	8.6	11.7	12.0					
	6.0	7.5	7.0	7.0	9.3	11.2	9.5	10.7	8.7					
	7.9	8.3	7.9	7.2	8.4	9.0	9.0	8.8	8.3					
	12.0	11.0	9.9	8.2	11.2	8.7	7.5	9.3	8.7					
	17.4	13.5	8.3	8.7	9.4	10.2	8.0	9.4	7.3					
42	2.0	11.6	10.5	11.7	10.6	12.2	16.9	17.6	18.9	7.8	2.5	3.2	16.4	13.1
	4.1	9.3	6.6	8.4	8.4	10.8	16.4	10.5	14.2					
	6.0	7.2	7.5	7.1	6.6	9.0	8.2	10.0	8.0					
	8.0	8.4	7.6	9.0	7.8	8.2	8.9	8.6	10.5					
	10.1	4.8	5.2	5.6	4.8	7.1	7.6	7.2	7.4					
	12.0	5.8	3.4	5.8	3.2	6.5	7.3	5.5	5.7					
43	1.3	8.4	13.9	8.3	6.5	14.4	13.4	18.8	12.8	6.2	2.7	2.8	14.1	11.2
	4.1	7.1	4.0	5.0	6.3	14.1	9.9	7.9	6.0					
	6.2	9.9	6.7	4.5	6.7	9.5	13.9	10.1	4.2					
	8.0	6.0	4.4	4.8	4.0	6.7	8.2	5.7	4.1					
	9.9	5.3	4.4	4.4	2.8	8.6	5.8	7.7	8.5					
	12.0	3.8	3.2	3.4	3.9	6.4	4.0	3.7	2.9					
44	1.1	82.9	39.3	45.8	41.9	42.5	85.1	52.3	62.2	18.6	5.6	8.8	36.5	27.7
	3.6	36.5	15.6	17.7	20.0	21.5	20.6	24.0	25.2					
	5.7	19.9	15.5	14.5	18.0	17.4	21.4	19.6	22.2					
	7.6	27.7	13.9	18.4	16.4	19.3	15.6	20.3	27.4					
	11.6	19.6	8.9	8.8	12.9	12.5	13.6	15.8	15.1					
45	1.2	5.5	9.3	7.0	6.5	5.9	11.1	11.5	13.1	7.6	1.8	4.8	11.4	6.6
	3.8	6.5	6.1	5.4	4.8	7.2	7.4	7.2	9.7					
	5.8	10.2	8.8	11.4	7.2	9.3	7.6	9.8	11.3					
	7.7	9.3	5.4	9.3	7.7	8.9	10.1	10.3	10.3					
	9.8	6.8	6.3	7.6	6.3	6.2	6.9	6.8	7.7					
	11.7	5.5	5.7	5.7	5.8	5.2	6.5	6.9	7.2					
46	1.5	46.6	27.2	30.0	39.3	32.7	23.8	30.5	43.1	13.7	4.6	7.0	25.7	18.7
	4.0	19.4	13.1	17.8	20.6	18.0	14.6	25.7	22.5					
	5.1	15.3	15.2	13.9	11.8	15.9	10.0	24.2	19.6					
	8.0	7.0	8.4	13.6	13.5	7.9	7.5	15.3	13.1					
	10.2	12.1	13.2	9.3	13.2	12.2	12.0	12.8	13.4					
	12.0	7.9	8.0	9.3	15.5	10.7	9.4	15.5	10.9					
47	1.5	13.1	13.1	1.3	16.4	16.4	10.9	13.1	16.4	7.0	1.9	3.1	10.9	7.8
	4.1	10.1	6.9	7.7	10.1	10.1	10.1	8.2	6.5					
	6.1	10.9	8.7	8.2	10.9	6.9	8.2	5.9	8.7					
	8.0	5.9	6.2	7.7	7.3	6.2	6.2	6.5	8.2					
	12.1	4.6	7.7	5.7	7.2	7.0	5.2	6.3	5.5					
	17.4	4.7	5.4	3.1	4.4	5.5	4.0	5.8	5.9					

Table C1 - Correlation coefficients for linear regression fitting of different corrosion rate measures as a function of environmental variable standard scores (Z).

Variable			Site Average						All Measurement Points					
No.	Name	NMLR	log (MLR)	CPR	log (CPR)	Pit Ratio	log (PR)	MLR	log (MLR)	CPR	log (CPR)	Pit Ratio	log (PR)
1	Site No.	47	NA	NA	NA	NA	NA	NA	NA	NA	NA	NA	NA
2	Soil	47	NA	NA	NA	NA	NA	NA	NA	NA	NA	NA	NA
3	Location	47	NA	NA	NA	NA	NA	NA	NA	NA	NA	NA	NA
4	Int Drainage	47	NA	NA	NA	NA	NA	NA	NA	NA	NA	NA	NA
5	Burial Depth	47	0.013	0.045	0.117	0.134	0.073	0.079	0.025	0.037	0.039	0.070	0.035	0.032
6	% Sand	34	0.275	0.352	0.132	0.181	0.222	0.260	0.218	0.304	0.026	0.092	0.134	0.203
7	% Silt	34	0.080	0.136	0.026	0.080	0.099	0.050	0.041	0.103	0.014	0.063	0.036	0.037
8	% Clay	34	0.289	0.338	0.152	0.167	0.201	0.294	0.249	0.306	0.022	0.066	0.143	0.232
9	% Colloid	34	0.297	0.344	0.187	0.208	0.176	0.270	0.253	0.311	0.042	0.089	0.128	0.213
10	% Suspen	34	0.239	0.301	0.311	0.277	0.472	0.561	0.194	0.253	0.203	0.190	0.286	0.437
11	Cond	47	0.511	0.556	0.370	0.373	0.376	0.397	0.445	0.483	0.186	0.244	0.220	0.288
12	Temp	47	0.368	0.340	0.088	0.087	0.339	0.357	0.319	0.286	0.049	0.071	0.213	0.247
13	Precip	47	0.158	0.053	0.182	0.089	0.037	0.043	0.091	0.017	0.055	0.007	0.033	0.027
14	Moistr Eq	47	0.315	0.423	0.270	0.323	0.215	0.307	0.291	0.390	0.103	0.185	0.151	0.240
15	Air Spc	45	0.213	0.338	0.054	0.019	0.411	0.478	0.201	0.323	0.048	0.004	0.253	0.357
16	Density	38	0.159	0.339	0.174	0.118	0.482	0.500	0.042	0.010	0.057	0.083	0.079	0.096
17	Vol Shrnkg	37	0.046	0.177	0.084	0.026	0.258	0.269	0.048	0.160	0.075	0.017	0.165	0.183
18	pH	47	0.142	0.117	0.141	0.096	0.068	0.045	0.092	0.054	0.009	0.001	0.048	0.062
19	Total Acid	41	0.570	0.442	0.403	0.375	0.126	0.240	0.499	0.401	0.205	0.234	0.103	0.187
20	[Na+K]	26	0.689	0.714	0.619	0.559	0.409	0.440	0.625	0.616	0.349	0.376	0.234	0.322
21	[Ca]	26	0.155	0.068	0.211	0.132	0.055	0.063	0.052	0.039	0.030	0.006	0.029	0.053
22	[Mg]	25	0.111	0.203	0.123	0.154	0.132	0.165	0.152	0.237	0.129	0.158	0.080	0.114
24	[HCO$_3$]	23	0.086	0.150	0.111	0.162	0.044	0.043	0.029	0.100	0.013	0.086	0.055	0.019
25	[Cl]	24	0.506	0.535	0.342	0.327	0.409	0.448	0.449	0.460	0.187	0.225	0.213	0.293
26	[SO$_4$]	26	0.618	0.671	0.615	0.572	0.380	0.404	0.561	0.587	0.349	0.391	0.197	0.271

Maximum Correlation Coefficient	0.689	0.714	0.619	0.572	0.482	0.561	0.625	0.616	0.349	0.391	0.286	0.437
Variable with Maximum Correlation	[Na]	[Na]	[Na]	[SO$_4$]	Den	% Sus	[Na]	[Na]	[Na]	[SO$_4$]	% Sus	% Sus

Table C2 - Correlation coefficients for linear regression fitting of different corrosion rate measures as a function of environmental variable standard scores (Z) for sites with chemical composition measurements.

Variable		Site Average					
No.	Name	NMLR	log (MLR)	CPR	log (CPR)	Pit Ratio	log (PR)
1	Site No.	NA	NA	NA	NA	NA	NA
2	Soil	NA	NA	NA	NA	NA	NA
3	Location	NA	NA	NA	NA	NA	NA
4	Int Drainage	NA	NA	NA	NA	NA	NA
5	Burial Depth	0.105	0.046	0.013	0.046	0.081	0.088
6	% Sand	0.367	0.410	0.437	0.423	0.020	0.038
7	% Silt	0.331	0.338	0.080	0.105	0.458	0.046
8	% Clay	0.539	0.576	0.230	0.200	0.384	0.425
9	% Colloid	0.500	0.560	0.230	0.222	0.365	0.407
10	% Suspen	0.284	0.284	0.424	0.435	0.596	0.636
11	Cond	0.613	0.689	0.594	0.562	0.407	0.438
12	Temp	0.413	0.391	0.159	0.106	0.432	0.439
0.03	Precip	0.152	0.026	0.217	0.084	0.102	0.106
0.25	Moistr Eq	0.190	0.331	0.157	0.210	0.344	0.382
15	Air Spc	0.022	0.083	0.180	0.190	0.338	0.345
16	Density	0.159	0.215	0.254	0.284	0.025	0.003
17	Vol Shrnkg	0.189	0.028	0.150	0.073	0.220	0.212
18	pH	0.057	0.016	0.034	0.072	0.032	0.012
19	Total Acid	0.700	0.678	0.603	0.524	0.353	0.398
20	[Na+K]	0.690	0.714	0.619	0.560	0.410	0.440
21	[Ca]	0.155	0.068	0.211	0.131	0.055	0.063
22	[Mg]	0.111	0.203	0.123	0.154	0.132	0.165
24	[HCO_3]	0.086	0.150	0.111	0.162	0.044	0.043
25	[Cl]	0.506	0.535	0.343	0.327	0.409	0.449
26	[SO_4]	0.618	0.671	0.615	0.572	0.380	0.404

Maximum Correlation Coefficient	0.700	0.714	0.619	0.572	0.596	0.636
Variable with Maximum Correlation	Total Acid	[Na]	[SO_4]	[SO_4]	% Sus	% Sus

Table C3 - Correlation coefficients for sequential multiple regression fits for mass loss rate.

Variable		1 Term		2 Terms		3 Terms		4 Terms		5 Terms		6 Terms	
No.	Name	N	R	N	R	N	R	N	R	N	R	N	R
5	Depth	47	0.045	47	0.558	47	0.685	23	0.854	20	0.918	20	0.943
6	% Sand	34	0.352	34	0.353	34	0.548	17	0.826	17	0.827	17	0.864
7	% Silt	34	0.136	34	0.278	34	0.508	17	0.821	17	0.821	17	0.858
8	% Clay	34	0.339	34	0.349	34	0.496	17	0.829	17	0.831	17	0.859
9	% Colloid	34	0.344	34	0.357	34	0.490	17	0.825	17	0.827	17	0.859
10	% Suspen	34	0.301	34	0.329	34	0.541	17	0.839	17	0.844	17	0.869
11	Cond	47	0.556		aaaaa								
12	Temp	47	0.340	47	0.674	bbbb							
13	Precip	47	0.053	47	0.606	47	0.684	23	0.845	20	0.921	20	0.948
14	Moistr Eq	47	0.423	47	0.573	47	0.692	23	0.844	20	0.920	20	0.951
15	Air Spc	45	0.338	45	0.568	45	0.664	23	0.849	20	0.923	20	0.942
16	Density	38	0.025	38	0.470	38	0.607	19	0.798	19	0.901	19	0.934
17	Vol Shrnkg	37	0.177	37	0.421	37	0.610	20	0.918		d		d
18	pH	47	0.117	47	0.576	47	0.680	23	0.874	20	0.923	20	0.944
19	Total Acid	41	0.442	41	0.597	41	0.671	18	0.788	18	0.845	18	0.876
20	[Na+K]	26	0.714	26	0.729	26	0.782	23	0.855	20	0.921	20	0.942
21	[Ca]	26	0.068	26	0.731	26	0.768	23	0.887	20	0.922	20	0.942
22	[Mg]	25	0.203	25	0.703	25	0.750	23	0.879	20	0.930	20	0.956
24	[HCO$_3$]	23	0.150	23	0.786	23	0.844		c		c		c
25	[Cl]	24	0.535	24	0.724	24	0.790	23	0.882	20	0.942		e
26	[SO$_4$]	26	0.671	26	0.711	26	0.792	23	0.849	20	0.918	20	0.942
	Max R		0.714		0.786		0.844		0.918		0.942		0.956

Table C4 - Correlation coefficients for sequentil multiple regression fits for the corrosion penetration rates.

Variable		1 Term		2 Terms		3 Terms		4 Terms		5 Terms		6 Terms	
No.	Name	N	R	N	R	N	R	N	R	N	R	N	R
5	Depth	47	0.134	47	0.381	23	0.649	20	0.747	20	0.844	20	0.872
6	% Sand	34	0.181	34	0.207	17	0.327	17	0.635	17	0.785	17	0.814
7	% Silt	34	0.080	34	0.088	17	0.378	17	0.660	17	0.784	17	0.815
8	% Clay	34	0.167	34	0.174	17	0.599	17	0.647	17	0.787	17	0.815
9	% Colloid	34	0.208	34	0.219	17	0.578	17	0.624	17	0.785	17	0.815
10	% Suspen	34	0.277	34	0.378	17	0.736	17	0.750	17	0.803	17	0.819
11	Cond	47	0.374	a	a	a	a	a					
12	Temp	47	0.087	47	0.391	23	0.647	20	0.755	20	0.843	20	0.861
13	Precip	47	0.089	47	0.387	23	0.647	20	0.784	20	0.841	20	0.891
14	Moistr Eq	47	0.323	47	0.399	23	0.647	20	0.774	20	0.842	20	0.882
15	Air Spc	45	0.019	45	0.418	23	0.684	20	0.788	20	0.860	20	e
16	Density	38	0.128	38	0.298	19	0.593	19	0.715	19	0.825	19	0.858
17	Vol Shrnkg	37	0.026	37	0.182	20	0.747		c		c		c
18	pH	47	0.096	47	0.382	23	0.646	20	0.832		d		d
19	Total Acid	41	0.375	41	0.403	18	0.398	18	0.519	18	0.718	18	0.765
20	[Na+K]	26	0.560	26	0.582	23	0.653	20	0.748	20	0.832	20	0.860
21	[Ca]	26	0.132	26	0.628	23	0.721	20	0.747	20	0.834	20	0.862
22	[Mg]	25	0.154	25	0.574	23	0.679	20	0.740	20	0.828	20	0.861
24	[HCO_3]	23	0.162	23	0.646	b	b	b	b				
25	[Cl]	24	0.327	24	0.573	23	0.666	20	0.750	20	0.834	20	0.865
26	[SO_4]	26	0.572	26	0.593	23	0.650	20	0.747	20	0.834	20	0.861
	Max R		0.572		0.646		0.747		0.832		0.860		0.891